VERY SOFT ORGANIC CLAY APPLIED FOR ROAD EMBANKMENT: MODELLING AND OPTIMISATION APPROACH

Very Soft Organic Clay Applied for Road Embankment:

Modelling and Optimisation Approach

DISSERTATION
Submitted in fulfillment of the requirements of
the Board for Doctorates of Delft University of Technology
and of the Academic Board of the UNESCO-IHE Institute for Water Education
for the Degree of DOCTOR
to be defended in public
on Tuesday March 11, 2008 at 10.00 hours
in Delft, the Netherlands

by

Cheevin LIMSIRI
born in Bangkok, Thailand

Master of Science in Hydraulic Engineering
IHE, the Netherlands

Taylor & Francis is an imprint of the Taylor & Francis Group, an informa business

Published by:
Taylor & Francis/Balkema
PO Box 447, 2300 AK Leiden, The Netherlands
e-mail: Pub.NL@tandf.co.uk
www.balkema.nl, www.taylorandfrancis.co.uk, www.crcpress.com

ISBN 978-0-415-38487-2 (Taylor & Francis Group)

Acknowledgement

I would like to express my sincere gratitude to my promoters; Prof. E. Schultz, PhD, MSc (UNESCO-IHE) and Prof. dr. J. D. Nieuwenhuis (Delft University of Technology) for contributing their knowledge, experience, and support throughout this work, and my supervisors; Ir. Piet Lubking and Drs. Gerard Kruse for their time, patience, the valuable guidance, encouragement, and continual support since the beginning of my study. I'm very grateful to Ir. A. H. Nooy v.d. Koff (Royal Boskalis Westminster nv) for his comments and reviewing the part of construction equipments and also Dr. ir. Dimitri P. Solomatine (UNESCO-IHE) for reviewing the optimisation analysis. Special thanks to Ms. Imke Deibel (Port of Rotterdam) and mw. ing. M. B. G. Ketelaars (Road and Hydraulic Engineering Institute, Rijkswaterstaat) for allowing me to collect samples at the sludge disposal site, Slufter and providing field data which was very useful for this thesis, Dr. Pichaid Varoonchotikul for his suggestions about Artificial Neural Networks and his encouragement.

I gratefully acknowledge Ir. Ronald de Heer and Dr. P. J. M. de Laat for the arrangement of my study, Ir. D. Boels for providing the RIJPING program, Dr. G. Greeuw for his suggestion on the FSCONBAG program and GeoDelft where this study was carried out. In addition, I wish to thank the staff in GeoDelft, UNESCO-IHE, colleagues and friends for their support.

My great appreciation is directed to my employer, Vongchavalitkul University, for her contribution in funding my study. Finally, I will always be indebted to my family, particularly my wife, Pimjai, for her sacrifice, understanding, and support throughout the period of my study, and my parents, Chatree and Anong, for their sacrifices throughout my life.

Summary

In this thesis strategies for using very soft organic clay as a fill material for road embankment construction are compared, and an optimisation scheme is presented. This thesis concerns optimisation of the use of clay in embankments. Also recommendations are presented based on a rational method for conditioning and emplacing very soft organic clay in a safe and economic way.

Very soft organic clay poses serious problems for embankment construction in deltaic areas. It is well known that this clay has very poor engineering properties; therefore it cannot be used as embankment fill material without improvement. Dewatering by evaporative drying, dewatering by horizontal sand drainage (clay-sand layers in a sandwich construction) and pre-loading (surcharging) are considered to be the simplest, most practical and most cost-effective methods for the improvement of the properties explored in this study.

Computer programs for simulation of evaporative drying, for simulation of consolidation, and for slope stability calculation were used to analyse clay behaviour. Field and laboratory testing programs for a trial embankment were set up in order to study the relations of parameters, to examine suitability and workability of the clay, and to verify the improvement methods. An optimisation technique was applied as a rational method for using very soft organic clay for road embankments leading to an economic construction. Optimisation models were developed for this particular problem. These models consist of an equipment selection model and a multi-objective optimisation model for clay-sand layered embankment constructions. The equipment selection model provides an optimum selection of the size and number of the earthmoving equipment fleet, which yields minimum equipment costs. A Genetic Algorithm (GA) was applied as optimisation technique to find the optimum solution. The results indicate that the equipment selection model and GA can efficiently be used to find the optimum, which is minimum equipment costs. The obtained equipment costs were used as parameters in the development of cost models that are part of the multi-objective optimisation of clay-sand layered embankments.

A special computer program was developed to integrate the embankment design to the GA optimisation process that is applied to find economic solutions in the multi-objective optimisation model. The calculation in the design part is performed using Artificial Neural Network (ANN) models. These models were developed from the results of the simulations in order to cope with complexity of the problem and to reduce computation time of the GA optimisation process. The ANN models for the calculation of the amount and duration of settlement, and of the stability were developed from the consolidation simulation program and the slope stability calculation program respectively. The model validation shows that with the aid of ANN models results are obtained that are almost identical to the results calculated by simulation programs.

The computer program has performed the optimisation process by using ε-constraints, or the so-called trade-off approach i.e. a situation in which the parameters can be changed as long as they contribute to a global optimum. The Pareto curve that illustrates this approach can be constructed from the set of optimum solutions. A proper decision for a clay-sand layered embankment construction can be obtained by the assistance of the trade-off curve. A sensitivity analysis was performed to study the sensitivity of the solution for design parameters, which include those related to embankment dimension, material properties, and material costs.

A clay-sand layered trial embankment was constructed by the Port of Rotterdam

Authority at the dredging sludge disposal site, The Slufter, to investigate the behaviour of such an embankment and to evaluate the potential of the reuse of very soft organic clay. Therefore the unpolluted Slufter clay was used as a fill material. Before the material was used in the construction evaporative drying in ripening fields and conditioning by mechanical equipment reduced the water content in the clay. The monitoring data were used to verify the evaporative drying simulation program and the ANN models. Comparison of the results shows that the water contents during drying obtained from the simulation program are higher than the monitoring data measured in The Slufter. Agitation of the clay in the ripening fields may be a main reason for this discrepancy. The ANN model for predicting the embankment stability shows a good agreement with the monitoring data. The ANN models for predicting the amount and duration of embankment settlements show some differences in relation to the monitoring data.

The results from the optimisation analysis and the trial embankment investigation were used together to develop a rational method for using very soft organic clay in road embankment constructions. Using the results of the procedure described above, a guideline was developed for conditioning and placing very soft organic clay as a fill material for road embankment.

Samenvatting

In deze dissertatie worden strategieën vergeleken voor het gebruik van zeer slappe, organische klei als ophoogmateriaal in wegconstructies; tevens wordt een optimalisatiemodel gepresenteerd. Deze dissertatie handelt over de economische toepassing van klei in ophogingen. Verder worden aanbevelingen gegeven om zeer slappe, organische klei op een veilige en economische manier te verwerken in de constructie.

Zeer slappe, organische klei veroorzaakt ernstige problemen bij de aanleg van ophogingen in deltagebieden. Het is bekend dat deze klei kwalitatief lage technische eigenschappen heeft; zonder maatregelen ter verbetering daarvan kan de klei niet als ophoogmateriaal worden gebruikt. Ontwatering door verdamping bij uitdroging, door horizontale zanddrains (klei-zandlagen in een sandwich-constructie) en voorbelasting (aanbrengen van een bovenbelasting) worden beschouwd als de meest eenvoudige, meest praktische en meest economische methoden ter verbetering van de eigenschappen die in deze studie zijn onderzocht.

Computerprogramma's voor de simulatie van verdamping bij uitdroging en van het consolidatiegedrag alsmede voor de berekening van de taludstabiliteit werden aangewend om het gedrag van de klei te analyseren. Terrein en laboratorium onderzoeken werden uitgevoerd ten behoeve van een proefterp waarin de betrekkingen tussen de parameters werden geanalyseerd, de toepasbaarheid en de verwerkbaarheid van de klei werden onderzocht en de grondverbeteringmethoden werden geverifieerd.

Voor het economische gebruik van zeer slappe organische klei ten behoeve van wegophogingen werd een rationele methode toegepast in de vorm van een optimalisatietechniek. Voor dit specifieke probleem werden optimalisatiemodellen ontwikkeld. Deze bestaan uit een materieel selectie model en een meervoudig optimalisatiemodel voor de constructie van een klei-zand-sandwich ophoging. Het materieel selectie model maakt een optimale keuze voor wat betreft de zwaarte of afmetingen en het aantal grondverzetmachines hetgeen leidt tot minimale materieelkosten. Voor deze optimalisatie werd een Genetische Algoritme (GA) techniek toegepast. De resultaten daarvan laten zien dat het materieel selectie model en de GA-techniek op een efficiënte manier leiden tot een optimale oplossing, dat wil zeggen minimale kosten. De gevonden materieelkosten worden gebruikt als parameters voor de ontwikkeling van kostenmodellen die deel uitmaken van de meervoudige optimalisatie van een klei-zand-sandwich ophoging.

Er werd een speciaal computerprogramma ontwikkeld om het ophogingontwerp te integreren in het GA-optimalisatie proces dat wordt toegepast om economische oplossingen te vinden in het meervoudige optimalisatiemodel. De berekening in het ontwerpdeel wordt gerealiseerd door middel van zogenoemde Artificial Neural Network (ANN)-modellen. Deze modellen werden ontwikkeld uit de resultaten van simulaties teneinde de complexiteit van het probleem te kunnen hanteren en de rekentijd van het GA-optimalisatie proces te reduceren. De ANN-modellen voor de berekening van de zettingsgrootte en -duur, en van de stabiliteit werden ontwikkeld uit respectievelijk het consolidatiesimulatieprogramma en het talud stabiliteitsprogramma. De model validatie laat zien dat de ANN-modellen resultaten opleveren die vrijwel gelijk zijn aan die welke met behulp van simulatieprogramma's zijn berekend. Het computerprogramma heeft het optimalisatieproces uitgevoerd door gebruik te maken van zogenoemde ε-constraints ofwel de zogenaamde uitwisselingbenadering dat wil zeggen een situatie waarin de variabelen kunnen worden gewijzigd zolang ze bijdragen tot een globaal optimum. De zogenoemde Pareto-curve die deze benadering in beeld brengt kan worden geconstrueerd uit de set van optimale oplossingen. Met behulp van deze curve kan voor een klei-zand-sandwich constructie de juiste oplossing worden bepaald. Om de gevoeligheid van de gevonden oplossing voor ontwerpparameters zoals ophogingsdimensies, materiaalparameters en -kosten te bestuderen is een gevoeligheidsanalyse uitgevoerd.

Door het Havenbedrijf Rotterdam werd een proefterp aangelegd in de vorm van een klei-zand-sandwich ophoging op het baggerspecie opslagterrein De Slufter om het gedrag van een dergelijke ophoging te onderzoeken en de mogelijkheid van hergebruik van zeer slappe, organische klei te evalueren. Daartoe werd de niet-verontreinigde Slufterklei gebruikt als ophoogmateriaal. Voordat het materiaal in de constructie werd gebruikt, werd het watergehalte in de klei verminderd door verdamping tijdens uitdroging in rijpingsvelden en door regelmatige omzetting van de klei door mechanische hulpmiddelen.

De meetgegevens van de proefterp werden gebruikt om het simulatieprogramma voor verdamping bij droging alsmede de ANN-modellen te verifiëren. Bij vergelijking van de resultaten blijkt dat de watergehalten tijdens het drogingproces, die door middel van het simulatieprogramma werden verkregen hoger zijn dan die werden gemeten in De Slufter. Dit verschil kan zijn veroorzaakt door de voortdurende omzetting van de klei in de rijpingsvelden.

Het ANN -model ter voorspelling van de taludstabiliteit vertoont goede overeenstemming met de meetgegevens. De ANN-modellen ter voorspelling van de grootte en de duur van de ophogingzakkingen laten enig verschil zien ten opzichte van de meetgegevens. De resultaten van de optimalisatieanalyse en de proefvakonderzoekingen werden samen gebruikt om een rationele methode te ontwikkelen voor de toepassing van zeer slappe organische klei in wegophogingen.

Uit de resultaten van de boven beschreven procedure werd een richtlijn ontwikkeld om zeer slappe, organische klei te verwerken als ophoogmateriaal in wegconstructies.

Contents

1
Introduction

1.1   Background

Clay may be used as a natural fill material for earthwork. Compacted clay is commonly used as a fill material in embankments, cores in dams and dikes, foundations for road layers, and liners on slopes. Engineering properties of clay depend on many factors such as water content, density and ageing. However, it is well known that behaviour of clay depends much on its water content and bulk density. A change in the water content of saturated clay is the greatest single cause of variation in its engineering properties. Shear strength, compressibility, and permeability are all, directly or indirectly, related to the water content of saturated clay soil.

In many delta areas clay is the raw material available in large quantities, either directly from the subsoil or from excavation at construction sites. In some areas such as the western part of the Netherlands and the central plain of Thailand this clay contains much organic matter. This organic clay is characterised by a very high water content, very low strength and very high compressibility. Normally, this clay is not used in engineering work. It has been regarded as useless or waste material and has always been eliminated by dumping in disposal areas or in the sea. At present, the amount of this material, as spoil from construction sites, is increasing due to development projects. Due to environmental concern and limited disposal area, handling the clay becomes a problem. Solving this problem efficiently is relevant. Re-use of very soft organic clay may be a way to solve at least part of the problem.

The very soft organic clay may be considered for re-use for road construction as embankment fill material. In delta areas the re-use of very soft organic clay is very attractive in view of the result of high price and scarcity of sand or good quality materials. However, due to the very poor engineering properties, the re-use cannot be done without soil improvement. It is considered that there are three conditions, which the clay must meet in order to be suitable for embankment purposes i.e. (Dennehy, 1978):

* *The ability to be excavated, transported, placed and constructed with normal equipment by normal methods;*
* *The ability to form embankments with stable side slopes;*
* *The capacity to have reduced settlements, which do not adversely affect the structural function.*

Obviously these objectives cannot be achieved with high water content clay, or very soft organic clay, for which the undrained shear strength is normally less than 5 kPa. These clay soils belong to the very soft clay group according to the American Society for Testing and Materials (ASTM), which defines clay that has an undrained shear strength of less than 12.5 kPa as very soft clay, and according to the Deutsches Institut für Normung e.V. (DIN), which uses a consistency index of less than 0.5 to define the very soft clay group (CUR, 1996). Hence soil improvement methods have to be used to improve the engineering properties of this clay.

The most important factor in determining the usefulness of very soft clay is its water content. Water content must be adapted to make it suitable for densification and to obtain suitable engineering properties. As the water content decreases the density of saturated clay raises, which means that its resistance to deformation and failure also increases and improvement of engineering properties occurs. Thus, reducing the water content or dewatering as much as feasible and practical is the most significant action in conditioning the clay for suitable embankment construction. Several methods have been used to treat unsuitable soil such as (Berggren, 1999):

* *Static methods*
  Static compaction rollers, Pre-loading, Vacuum, Ground water lowering, Compaction grouting;
* *Dewatering methods*
  Electro-osmosis, Vertical and horizontal draining, Evaporative drying;
* *Dynamic methods*
  Falling weight, Blasting, Vibrating plate/roller;
* *Chemical methods*
  Lime/cement columns, Grouting, Shallow stabilization;
* *Thermal methods*
  Freezing, Heating;
* *Interaction methods*
  Reinforcement with geo-textiles, Grid and nails.

Among these methods, dewatering by evaporative drying, dewatering by horizontal sand drainage and the accelerated consolidation process by pre-loading are considered to be the simplest, most practical and most cost-effective methods. These methods have been used in many dams, highways and land reclamation fills to reduce excess pore-pressures, moisture contents and to increase the undrained shear strength of the fill (Thomas, 1993). These methods are considered potential methods for improving very soft organic clay to become suitable fill material for road embankment, addressed in this study.

Dewatering by evaporative drying, vertical and horizontal drainage with preloading are normally used to dewater the clay fill. Krizek et al. (1973) investigated several techniques of dewatering, and suggested that evaporation is probably one of the most useful and economical means for dewatering thin lifts of soil materials (Krizek et al., 1978). During dewatering by evaporative drying, wet clay will be left at a stockpile and allowed to dry until it is suitable for using as fill material. The water in the pores of the clay evaporates improving the strength of the clay. Volumetric shrinkage during desiccation usually causes vertical shrinkage cracks. This effect tends to accelerate the desiccation processes, and make drying behaviour complex (Krizek et al., 1978; Fujiyasu et al., 2000). Surface trenching and regular mixing have always been used to accelerate the drying rate of wet clay. Because of the simple operations and the relatively low costs, evaporative drying is a method, which is widely used in clay embankment construction. It will be clear that drying wet clay by evaporation processes depends much on the weather conditions. In unfavourable weather, drying may be complex.

Adding chemical agents is another method that can dewater and improve engineering properties of very soft clay but the efficiency of the improvement depends on many factors. These factors are soil gradation, types of clay minerals, amount of organic matter, kind and amount of chemical agents, moisture content, mixing, amount of compaction, curing time and curing temperature (Mateos, 1964). Thus, using this method needs a specific study, special equipment and also very careful operation, which lead to additional cost and time. Therefore, the chemical method has a limited place in this study.

When very wet soft organic clay is applied as a fill material for road embankment construction, consolidation and stability are problematic. To prevent such problems, the water in the clay has to be removed as fast as possible. The time it can take to remove the water or consolidate relatively thick compressible layers can be long. Any technique, which decreases the drainage path length through the clay, would be helpful in decreasing the time required for compaction. Making layered clay-sand embankment is one of the construction methods which speed up consolidation and improve strength more rapidly as compared to a full clay fill owing to the shorter drainage path provided by the sand layers (Inada et al., 1978; Ostlid, 1981; and Tan et al., 1992). Thus, this method may be recommended too when using very soft clay as fill material for embankment. The works of Gibson and Shefford (1968), Koppula and Morgenstern (1972), Lee et al. (1987), Karunaratne et al. (1990), and Tan et al. (1992) show that engineering behaviour and workability of layered clay-sand embankments are dependent on many parameters such as the thickness, permeability, and the relative length of the flow path for two materials and there seems to be no consistent rationale for their design and construction. Thus, use of this method still needs a further study.

Above methods have been applied to construct embankments of wet clay, and the result showed the potential of applying these methods in embankment construction with wet or high water content fill material. However, it is not clear which method to use in which circumstances. How to use very soft clay economically as a fill material is not clear. Questions concern:
* How large a stockpile will be used;
* How long it will take for drying;
* What is the proper thickness of clay for drying;
* What is the proper thickness of clay and sand in a layered clay-sand embankment;
* Comparison of drying and the accelerated consolidation process: which is the most suitable method for embankment construction;
* How to combine these improvement methods and yield the most beneficial construction.

Simulation computer programs, a laboratory testing program, large-scale models and trial embankments were used in order to study the relations of parameters, examine suitability and workability, and verify the improvement methods. An optimisation technique has been applied in the development of a rational method for selecting and applying the most suitable improvement method. The format of this study has resulted in a selection path asking for many technological and managerial decisions at several stages. The ultimate goal in such decisions is to achieve insight in factors that determine the maximum cost-effectiveness for road embankment construction, and this goal can be achieved by optimisation. Optimisation is the act of obtaining the best result under given circumstances. Because there is no single method available for solving all optimisation problems efficiently, for this particular problem an optimisation model has been developed using a suitable optimisation method. The optimisation model was developed based on computer simulation programs, test results, and construction cost data. Economic costs are considered an effective means for optimisation since the parameters in the cost can cover all activities in construction and qualities of the resulting product. The objective of the optimisation is either to maximise cost-effectiveness or minimise construction cost in road embankment construction. This approach offers the potential for a better and more appropriate use of resources, and reduction in waste.

Due to the complexity of this problem a direct search method that is widely used and is effective for analysing optimisation-engineering problems is proposed as optimisation technique. Besides the criteria of maximum cost-effectiveness, suitability and workability have been established in order to guide and control the constructions.

Optimisation for the use of very soft organic clay as a fill material for road embankment economically and safely was the goal of this study.

1.2   Research objective

The objective of this research was to develop an optimisation schedule enabling the development of recommendations or guidelines for using very soft organic clay as a fill material for road embankment construction in a safe and economic way.

1.3   Scope of the research and desired research results

In order to achieve the above-mentioned objective, a systematic study has been carried out. The study comprised literature review, laboratory tests, cost analysis, optimisation analysis, and field-testing. The research program has been implemented as follows:
* *Literature study*
  Literature and specifications concerning soil improvement, evaporative drying, consolidation accelerated by horizontal sand drainage, suitability and workability of clay for embankment as well as settlement and stability of clay embankment, optimisation, clay embankment construction and cost have been studied. The models, the relations of the parameters, and simulation computer programs, which concern evaporative drying, consolidation accelerated by horizontal sand drainage, consolidation and stability of clay embankment have been studied and selected for use as guideline for setting the test program and optimisation models;
* *Laboratory tests*
  A laboratory-testing program has been set up and was conducted for determining basic properties, engineering properties, and workability of improved clay;
* *Large-scale model tests*
  A trial embankment was constructed in order to investigate the behaviour of dried clayey dredging sludge, the possibility and efficiency of the improvement methods have been analysed, and verified with the computer programs which have been used to simulate clay behaviour during the improvement. The construction costs as well as the behaviour of the trial embankment have been investigated;
* *In-situ tests*
  An in-situ testing program has been implemented for monitoring the development and behaviour of the improved clay;
* *Cost analysis*
  The cost data of the embankment construction equipment that had to be used in the improvement procedure and embankment construction have been collected. The investment and operation costs equipment could be determined from these data. Related costs in construction cost for each improvement method have been collected as well. The result of the cost analysis is a significant part of the optimisation analysis;
* *Optimisation analysis*
  Optimisation models have been developed for the optimisation analysis. The results of the optimisation analysis have been used for decision-making. The best solution for using very soft organic clay for road embankment such as construction method, proper water content of clay after drying, and details of embankment structure could be obtained from the analysis. Equipment selection models and construction cost models have been developed based on the results of the simulation program, large-scale model tests, and cost analysis. The optimisation analysis scheme is shown in Figure 1.1. The objective of the optimisation is the

maximum cost-effectiveness in construction or minimum construction cost. The direct search method and multi-objective optimisation were used to find the global optimisation result;

* *Development of a rational method and recommendation of using very soft organic clay*

Recommendations and a rational method for using very soft organic clay as a fill material for road embankment construction efficiently and economically have been developed based on the results of the optimisation analysis. Suitability and workability of very soft organic clay during the improvement procedure have also been assessed.

1.4   Research methodology

The first task in this study was to gain and examine the already available knowledge, related to the objective of this study. This work required an extensive literature review in the areas outlined in the previous section. Parameters that are relevant for the objective of this study were determined. A laboratory test program was set up to determine and evaluate the basic properties of the clay. Simulation programs have been used to determine the relationship of the relevant parameters, which were used for developing mathematical models. Large-scale model tests were done to investigate and verify the results from the simulation programs. Workability and the problems that can occur during construction were also observed from these large-scale models. At the same time a cost analysis was carried out by collecting the data of the embankment, construction equipment, and related costs. The equipment and construction costs have been determined and were used in the mathematical models.

1.5   Thesis outline

This thesis consists of 7 chapters described in very brief terms below. A diagram of the relationship between the chapters is shown in Figure 1.2.

The first chapter describes problem, objective, scope, and methodology of this research.

Chapter 2 gives an overview of the basic concepts, terminology and related work on the problems that were the focus in this study.

In Chapter 3 the equipment for embankment construction and cost analysis that will be used in the optimisation models are described.

In Chapter 4 Genetic algorithms (GAs), which is the optimisation technique that is used in the optimisation analysis of this study, is described as well as Artificial Neural Networks (ANNs) which have been used as a tool to find the relationship between input and output parameters of the simulation programs.

In Chapter 5 the optimisation models, the optimisation analysis, and the sensitivity analysis are presented.

In Chapter 6 the field and laboratory tests, and the proposed rational method developed from the optimisation results and field investigation are presented.

In Chapter 7 the main results of the research are summarised.

Very soft organic clay applied for road embankment

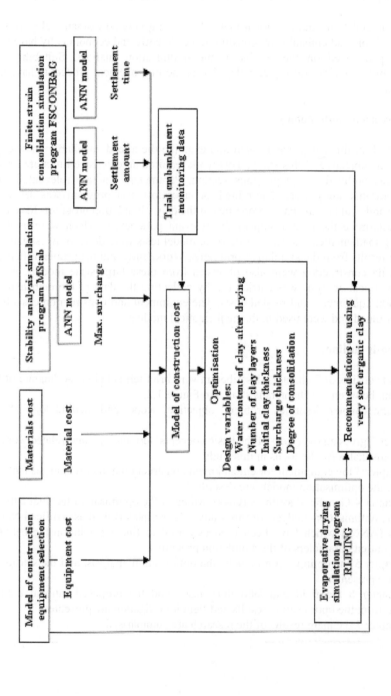

Figure 1.1: Optimisation analysis scheme

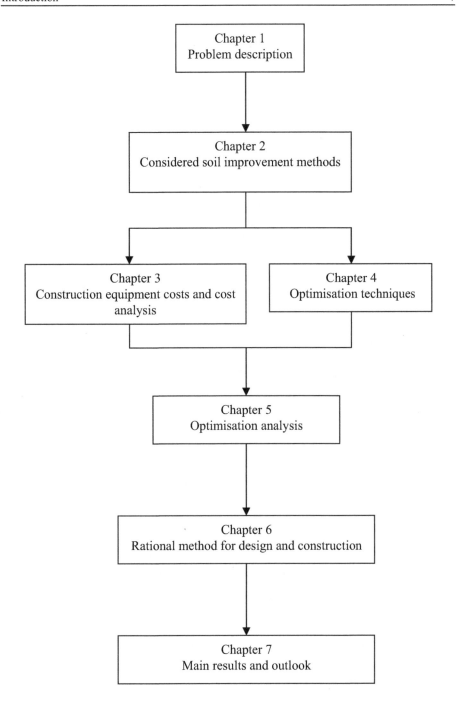

Figure 1.2: Thesis outline

2
Soil improvement methods considered

This chapter gives an overview of the basic concepts, terminology and related work on the problems that have been the focus of this study.

2.1   Very soft organic clay

The behaviour of a soil changes with the water content especially in clay soils. At high water contents soils are in suspension, with flow properties of liquids. As the water content is gradually reduced, the flow properties of clay soils change to the non-Newtonian flow of pastelike materials. On further drying the clay can be moulded. At this water content it is plastic. As the water content decreases still further the plasticity is lost, the soil becomes harder to work and gradually takes on the properties of a solid at low water contents.

The ability of material to change shape continuously under the influence of an applied stress, and to retain the new shape on removal of the stress is plasticity. This distinguishes it from an elastic material, which regains its original shape on removal of a stress, and from a liquid, which does not retain its own shape. Only the very small particles of soils of the right mineralogy, clays, and to some extent silt exhibit plastic behaviour at low stresses. The deformation behaviour of a soil at given water content is called its consistency. Consistency is the resistance to flow of the soil and also an indication of its deformation behaviour. Consistency is related to the force of attraction between individual particles or aggregates of these particles and geometrical interferences. Different soils have different consistency at different water contents, and the specification of this condition gives some information about the type of soil material.

The range of water content over which a soil exhibits plastic behaviour at low stresses is defined as that between the plastic limit (w_P) and liquid limit (w_L). The plastic limit is that water content below which the soil is considered not plastic when it is worked, and crumbles on application of pressure. At the liquid limit the change is from plastic to flow behaviour. These measurements of plasticity are based on the work of Atterberg, thus the limits are called Atterberg limits (Budhu, 2000). Measurements of plasticity are made on remoulded soil. Plasticity, in conjunction with particle-size distribution, is widely used in evaluation and classification of soil for engineering purposes. The liquid and plastic limits depend on both the type and amount of cohesive particles present in the soil. The difference between these two values is called plasticity index (I_P). Compressibility of remoulded soil increases with increasing water content at the plastic limit, whereas strength of soil decreases under the same conditions. The shrinkage limit is the moisture content at which further decreases in moisture content do not cause further shrinkage. This limit is seldom used in routine engineering work (Rollings and Rollings, 1996).

The state of consistency of a natural soil can be established through a relationship termed the liquidity index (Budhu, 2000):

$$I_L = \frac{w_N - w_P}{w_L - w_P} = \frac{w_N - w_P}{I_P}$$

(2.1)

where I_L = the liquidity index
$\quad\quad\quad w_N$ = the natural moisture or in situ water content (%)
$\quad\quad\quad w_P$ = the plastic limit (%)
$\quad\quad\quad w_L$ = the liquid limit (%)
$\quad\quad\quad I_P$ = the plasticity index

Another relationship is the index of consistency, defined as (Bowles, 1979):

$$I_C = \frac{w_L - w_N}{I_P}$$

(2.2)

or:

$$I_C = 1 - I_L$$

(2.3)

Both equations give an index value between 0 and 1 when the water content is between w_P and w_L.

Strength of clay can be classified as very soft, soft, firm, stiff, hard, and very hard. Very soft organic clay, which will be treated in this study, normally has an undrained shear strength of less than 5 kPa, liquid limit (w_L) very high, and natural water content (water content is defined as "the ratio of weight of water in the soil to weight of the solid matter") at the liquid limit, thus these clay soils can be classified as very soft clay. The classification can be seen in Figure 2.1 where the correlation between undrained shear strength and consistency for remoulded clay developed by Wroth and Wood (CUR, 1996) and the correlation of the Atterberg limits and the prevailing water content on undrained shear strength are shown.

The presence of large amounts of organic matter in soils is usually undesirable from an engineering standpoint. The organic matter may cause high plasticity, high shrinkage, high compressibility, low permeability, and low strength (Mitchell, 1976). Organic matter in the soil is derived from mainly plant (or animal) remains. Organic soil may occur at depths, where the normal processes of decomposition are arrested due to lack of oxygen (Scott, 1980). Particulate organic matter in soils is complex both chemically and physically and varies with age, origin, and environment. The specific properties of the fine colloidal particles vary greatly depending upon the parent material, climate, and stage of decomposition. It occurs mostly as: carbohydrates, proteins, fats, resins and waxes, hydrocarbon, and carbon. Cellulose ($C_6H_{10}O_5$) is the main organic constituent of soil. Organic particles may range down to 0.1 μm in size. Organic materials have properties, which can be undesirable in engineering works. These properties are summarised below:

*   Organic material can absorb large quantities of water. Large volume changes occur as the result of water expulsion when pressure is applied to the soil;
*   The material has very low shear strength and will adversely affect the strength of any soil of which it forms a considerable part;
*   The presence of organic matter generally inhibits the setting of cementing substances. Highly organic soils can hardly be stabilised with cement.

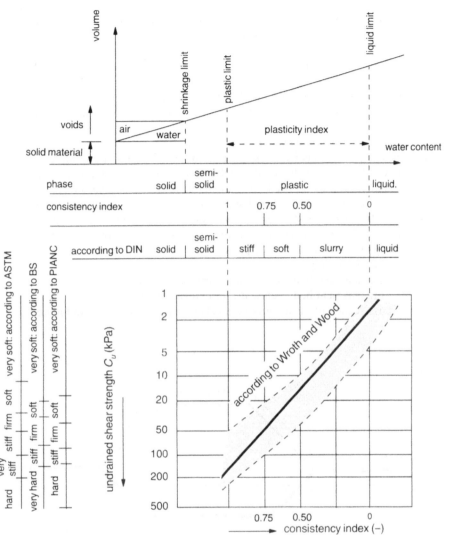

Figure 2.1: Relationship between undrained shear strength and the consistency index (CUR, 1996)

2.2 Embankments

A road embankment is a construction of earth or rock material bringing the road surface at desired elevation and to transmit the traffic load to the natural ground without causing unacceptable deformations and in particular differential settlements. Embankments can be applied for highways, airfields, and railways. Other applications are to raise the ground level that can be used to retain water or dredged materials as dikes, dams, and banks or to reduce noise as noise barriers.

Embankments are constructed of materials that usually consist of soil, but may also include aggregate, rock, or crushed stone-like material. Embankments can be constructed on natural firm ground or replacing the upper soft soil by a fill of the better strength soil compacted to a specific requirement. The position of the groundwater table affects the behaviour of the embankment during its lifetime. Since different soil types behave differently under saturation, the types of soil to be used in the embankments and the compaction requirements are also governed by the groundwater level with respect to the embankment.

2.2.1 Types of embankments

There are two main types of embankments, non-structural embankments and structural embankments, which are used to transmit the traffic load to the natural ground (CUR, 1996):

*   *Non-structural embankments*
     These types of embankments are used to raise the natural ground level so that the effect of ground water to the embankment surface is minimized. The elevated ground level can be used for recreation, traffic noise barrier and other similar purposes. These types of embankments do not carry surcharge loads. However, they should be constructed to carry their own weight and avoid some adverse effects of water either from within the ground or from rain. These types of embankments consist of three parts:
     1.   Surface lining;
     2.   Embankment body above the groundwater table;
     3.   Embankment body below the groundwater table.
     A cross-section of raising ground level embankment is shown in Figure 2.2.

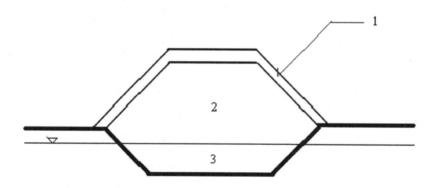

1 – Surface lining
2 – Embankment body above the groundwater level
3 – Embankment body below the groundwater level

Figure 2.2: Cross-section of raising ground level embankment

*   *Embankments to transmit the traffic load to the natural ground*
     The purpose of these types of embankments is to carry traffic load such as highways and railways. The requirement to withstand diverse effects of combined traffic load and changes in water content variation is stringent when compared to embankments merely intended to raise the ground level. Therefore, unlike the latter embankments, these types of embankments require more attention for the material used in their construction. These types of embankments have to carry surcharge

loads in addition to their own weight and transmit the load to the natural ground. Each layer of the structure should be able to resist the effect of change in moisture content. Such embankments consist in general of five parts:

1. *Base course*
   This is the foundation of the surface layer, which distributes the imposed traffic load to the sub-base. The layer is usually constructed from permeable crushed rock or structural materials;

2. *Sub-base course*
   This is the layer supporting and further distributing the loads from the base course to sub-grade. The sub-base course is usually constructed from very stiff soil such as highly compacted sand on well-compacted sub-grade;

3. *The sub-grade above the terrain*
   This layer must not deform in short and long term;

4. *Sub-grade below the terrain*
   When the existing soil is too soft and too compressible to support the overlying and traffic load, it is removed and replaced by well-compacted suitable soil. The material in this layer is almost always under water, therefore the compaction and the soil type should prevent failure in the saturation condition;

5. *Surface liner*
   This is the same as that of raising ground level embankments.

The cross-section of an embankment for roads is shown in Figure 2.3.

This study focuses on embankments for roads. The application of treated very soft organic clay in the sub-grade layer above the groundwater table for such embankments has been studied.

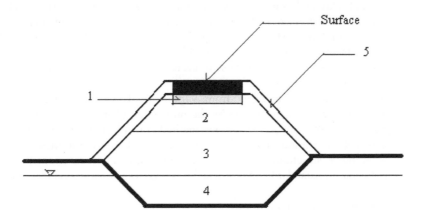

1 – Base course
2 – Sub-base course
3 – Selected fill sub-grade above the groundwater table
4 – Selected fill sub-grade below the groundwater table
5 – Surface liner

Figure 2.3: Cross-section of an embankment for roads

2.2.2   Materials used in embankments

Selection of materials to be used in embankment construction is mainly based on the construction technology and the availability of the materials at reasonable cost. The limit on the knowledge of material properties also affects the use of materials. The construction materials can be classified into two categories, primary and secondary materials. Primary materials are natural fill materials that be excavated specifically for earthworks. The term "Secondary materials" generally refers to that category of processed or unprocessed residues used in earthworks as an alternative to primary materials:

*   *Primary materials*
     Primary materials used in embankment construction are natural fill materials such as crushed rock or gravel in the base course, sand in the sub-base and sand or clay in the sub-grade course. Although sand is a good quality material the use of sand can be costly in some areas where clay is prevailing such as in many large deltaic areas;
*   *Secondary materials*
     Extraction of primary materials has become more problematical due to dwindling of primary raw materials. The use of secondary materials in embankment construction has increased as the construction methods improved through time. Besides the primary construction materials, other natural materials and waste materials from industries can be applied in road embankments:
     *   *Natural materials*
         These include naturally occurring materials such as cohesive soils like soft to very soft clay and non-cohesive soils like silty sand. Cohesive soils and non-cohesive fine-grained soils did not fulfil the specification requirements in natural condition. However, at present these materials can be used in embankment construction due to the improvement of the design and construction methods. But the use of these materials needs precautions;
     *   *Industrial waste materials*
         In recent years, researchers from various fields have attempted to solve environmental problems posed by the production of industrial wastes. Gidley and Sack (1984) suggested various methods of utilisation of these wastes in construction. Various researchers studied potential use for these materials including a structural fill. However, a specific study is needed for any particular non-standard material when it is used as a construction material.

2.2.3   Moisture regime in embankments

Slope of embankments tend to be finished off with a surface covering called liner. This surface covering is commonly constructed with clayey soil. The clayey soil in the unsaturated zone is subject to swelling and shrinkage resulting in fractures. Repeated extreme wetting and drying conditions promote further fracturing. Consequently, surface clay is very permeable. As water infiltrates the embankment, instability of the embankment can occur due to slope softening (Day and Axten, 1990; Kyfor and Gemme, 1994). To prevent such action the deeper layers have to be impermeable. Intense compaction in the deeper layers can reduce the water infiltration. The water content of a compacted soil in a construction will tend to an equilibrium range. If soil is compacted at too high a water content compared to this equilibrium range, a permanent volume reduction results. This shrinkage will result in permanent wide cracks. Thus, clays are best compacted at water contents at approximately a longer-term equilibrium water content related to the average negative pore pressure at the location in the construction. In the Netherlands a proper water content of clay liners during compaction is less than the water content at $I_C = 0.75$ and for clay in the core of embankments $I_C = 0.6$. For temperate climate conditions such as France's summer $I_C = 0.8$ (Kruse and Nieuwenhuis, 1998).

2.3   Evaluation of clayey fill materials

Fill material is required for the construction of the subgrade of embankments. In order to construct an embankment it requires the excavation, transportation, placement and compaction of this fill. Workability including suitability of the fill material has to be considered in the construction, which is important to estimate cost and time required for construction. The ability of the construction equipment to operate effectively and the effect of plant operations on the quality of fill material affect the efficiency and costs of embankment construction.

Clay is one of the fill materials used in embankments and water content and density tend to be a determinant. In order to achieve the functions of embankment specification control of quality of material is required. The specification is aimed to control:
*   Stability and settlement requirements (usually in the case of high embankments);
*   Impermeability (for longer term stability);
*   Potential difficulties in plant operation.

Quality control of the material is necessary to meet the requirement. Direct measurements of the relevant engineering properties of the soil or indirect types of measurement, which correlate with the engineering property, are used. Very soft organic clay has insufficient quality for use as fill material for road construction if it is not improved.

In this study, a method of specification and control for use of improved soft clay will be outlined.

2.3.1   Construction demands

The quality of the soil being excavated can vary as a result of changes in ground condition, in soil type or in the weather due to the sensitivity for changes in water content. Control of the quality of the material is necessary to qualify the suitability of the material. Allowance has to be made, however, for the effects of weather conditions on the quality of the soil during the construction process. The suitability criterion is determined from considerations of trafficability by earthmoving plants, embankment stability, and embankment settlement:
*   *Trafficability*
     Plant operation is a critical criterion in embankment construction. It is directly linked to construction costs. Trafficability problems occur with weak soil materials. The wheels of plant may sink where the soil is over-stressed, causing deep ruts as a result of excessive plastic deformation of the soil. Slurrying of the soil may occur due to kneading of tyres. Considerable additional power will be needed to move the plant forward against the soil drag. The study by Dennehy (1978) stated that ruts of 275 mm represented the trafficability limit of the fully loaded medium to heavy scrapers (capacity greater than 16 m^3) such scrapers can operate without assistance on fill and at rut depths less than 200 mm. This indicates that ruts of about 200 mm to 275 mm would define the area of marginal

trafficability. These rutting limits can be related approximately to the required undrained shear strength of soils (C_u) for the two tyre pressure ranges as shown in Table 2.1.

Table 2.1. Traficability criteria of rubber-tyre vehicles (scraper type) on cohesive soils

Scraper type	Tyre pressure (kN/m^2)	Shear strength (kN/m^2)	
		Unsuitable	Suitable
Light to medium	240 - 310	<40	<60
Medium to heavy	340 - 380	<60	<80

Arrowsmith (1978) suggested rather lower figures, which were 35 kN/m^2 for Caterpillar tractors and scrapers, and 50 kN/m^2 for large rubber-tyre scrapers. This discrepancy may be due to various factors among other type of C_u determination;
*   *Embankment stability*
    Stability of embankments is considered either as a short-term stability or a long-term stability problem. In the short-term stability problem, the stability is usually determined using total stress analysis using unconsolidated-undrained shear strength parameters. In the long-term stability problem, an effective stress analysis using effective shear strength parameters is carried out. The influence of moisture content and density of the fill and the change over time of the material on slope stability and settlement has to be evaluated. Dennehy (1978) indicated that for most cohesive soils, except those of low plasticity, which are placed at undrained shear strengths of between 40 kN/m^2 and 60 kN/m^2 no problems, neither self-settlement nor slope stability of embankments placed on hard foundation arose. At these strengths embankments with side slopes of 1:1.5 or 1:2 will be stable at heights between 9 m and 14 m. If the embankment were to be built on weak foundation a more detailed analysis would be required;
*   *Embankment settlement*
    Settlement is a function of embankment height, density, moisture content, air void after compaction, and the compressibility characteristics of the soil. The settlement may affect the functioning of embankments. In order to construct satisfactory embankments it is necessary to reduce the void ratio of the compacted fill to a value, which would ensure that only minor or at least acceptable additional settlements would take place after the completion of the works. Within embankments modest differential settlements can be minimised if the fill is placed in layers of even thickness and ruts are progressively filled before the placement of the next layer.

2.3.2 Existing specifications

Methods of specification and control, which are normally used in road construction, can be summarised as follows:
*   *Moisture content and plastic limit*
    The limit of acceptability of a material may be specified in terms of moisture content. Specification of the Ministry of Transport of the UK in 1963 prescribed that cohesive soil at a moisture content within plus or minus 2% of its plastic limit can be used as fill material. The method worked reasonably well for clays of low plasticity with a low stone content. For more plastic clays this Specification declared soil "unsuitable", which was quite strong enough to form a stable

embankment (Arrowsmith, 1978). From experience on the construction of the M6 in the UK between 1961 and 1963 it was suggested that the upper limit should be +2% for clays of low plasticity, +4% for medium plasticity and +6% for high plasticity with plasticity defined by the plasticity index. The Ministry of Transport of the UK used the comparison of moisture content with plastic limit as a means of assessment of the suitability of clay fill in highway embankments in 1969. The upper limit of 1.2 multiplied by the plastic limit was derived from the experience of the M6 motorway embankment construction (Arrowsmith, 1978). However, there were substantial difficulties associated with the use of the ratio of moisture to plastic limit as method of control, especially when undrained shear strength was the engineering property of interest. The poor correlation between the ratio and the undrained shear strength was mentioned by Parsons (1978). One reason suggested for the poor correlation is the low accuracy of the plastic limit test and the ensuing variability of the results;

*   *Moisture content and liquid limit*

Because of the increased accuracy, which can be obtained from the cone penetration liquid limit test, the liquid limit was suggested as a preferable alternative to the use of the ratio of moisture content to plastic limit as a measure of the quality of the soil. Dennehy (1978) showed that the liquid limit is a more accurate index of strength than the plastic limit. Powell suggested the likely limit for suitability would be about 0.5 times the liquid limit (Dumbleton and Burford, 1978). However, Black and Lister (1978) indicated that the method is not universally applicable;

*   *Wetness index*

A further extension relating suitability with water content is the inclusion of the so-called optimum moisture content. Optimum moisture content is determined as water content at which a maximum value of dry density is obtained in the compaction test. The use of a wetness index was proposed by Dohaney and Forde (1978), which involves the measurement of two parameters as well as the moisture content.

$$Wetness\ index = \frac{w_L - natural\ moisture\ content}{w_L - optimum\ moisture\ content}$$

These parameters are liquid limit (w_L) and the optimum moisture content determined in a compaction test.

*   *Consistency index*

A consistency index:

$$Consistency\ index = \frac{w_L - natural\ moisture\ content}{w_L - w_P}$$

For the consistency index the liquid limit and the plastic limit are parameters as well as the moisture content.

*   *Moisture condition test*

Transport and Road Research Laboratory (TRRL) in UK introduced the moisture condition test in 1976 in an attempt to produce an improved method for establishing earthworks suitability (Green and Hawkins, 1987). The test basically consists of determining the compactive effort necessary, in terms of the number of blows of a rammer, to fully compact a sample of soil. The details of the test are given in Parsons (1976). The test is relatively rapid and is reported to be applicable over a wide range of soil types. The resulting moisture condition value (MCV) is shown to correlate well with the undrained shear strength, moisture content and California Bearing Ratio (CBR). The classification of soil possibly determines the relation between MCV and moisture content as demonstrated by Parsons (1978). This method was included in the Specification for Highway Works and associated Notes for Guidance 1986, Department of Transport, UK;

*   *Direct measurements of strength*
    The direct measurement of undrained shear strength (C_u) can be done by the unconfined compression test, triaxial test, or vane test. The type of soil and the methods of sampling and testing can have an influence and give rise to different results. It appears that direct measurement of strength could work well and is generally preferable in homogeneous clay soils (Parsons, 1978). Dennehy (1978) suggested that the most appropriate way of specifying the suitability of clay materials for fill purposes is an undrained shear strength (C_u) basis. Embankment stability and trafficability can be specified by certain values of undrained shear strength (C_u). Charles and Watts (2001) stated that a clay fill can be specified in terms of undrained shear strength (C_u) and there is a range of values of undrained shear strength (C_u), typically 50 kPa < C_u < 120 kPa, which will generally be acceptable. This does not include the behaviour on longer term and changing environment;

*   *Processibility classification*
    The suitability and workability can be assessed by systematic classification of materials and boundary conditions. Soil type, moisture content condition and weather characteristics during processing are the data used for the classification. The result from the classification leads to the determination of the suitability of the material and the method of construction.

The use of the indexes as method of control is problematic due to sampling discrepancies and inaccuracy of test results.

2.4   Wet fills

It has been common practice when preparing specifications for constructions to specify the best material available. However, nowadays specifications of the widest range of materials, which can perform satisfactorily in the finished works, are requested due to economic and environmental reasons. Hence, for embankment construction the use of wet clay, which has often been classed as unsuitable, gets into the picture. Results of many studies have shown the potential of the use of wet clay as fill material for embankment and reclamation if drainage layers are included (Boman and Broms, 1978; Grace and Green, 1978; Inada et al., 1978; Ostlid, 1981; and Lee et al., 1987). Drainage layers can increase the rate of dissipation of excess pore pressure, hence increase the shear strength of wet clay during construction and decrease the settlement after finished construction in a shorter period. The result of Boman and Broms (1978) indicated that soft clay can be used economically in embankments instead of sand, and drainage layers can reduce the time required for consolidation of the clay considerably. At a degree of consolidation of 60% or higher, most clay materials have enough strength to be stable in embankment construction. Settlement during further consolidation would require a rather flexible road structure however to maintain integrety of the road. Grace and Green (1978) described construction problems when using wet fill as:

*   The use of wet fill requires placing and compacting techniques different from those

normally employed. The wet fill is unable to support construction traffic and suitable haul roads have to be provided. Extra large track bulldozers can affect spreading and one or two passes of the bulldozer can achieve a considerable reduction in percentage of air voids;

*   The bearing capacity of wet fill is very low. Thus 1.8 - 2.4 m of suitable fill should be placed to obtain a firm surface before any base or sub-base is laid;
*   It is common practice to select relatively flat areas for the wet fills and they are normally contained by construction of bunds of suitable fill placed on each side.

Inada et al. (1978) investigated the long-term stability of a high embankment with soft volcanic soil as a fill material and found that the embankment with many drainage layers became stable at an early stage and remained in a good condition for a long time.

The temporary surcharge loading of a clay fill with drainage layers accelerates consolidation. However, a surcharge of fill on top of an embankment of wet clay fill might well cause instability if placed without proper stability evaluation.

All of these works concerned only wet clay without organic matter and do not mention efficient and economical design and construction. In this study the unaddressed elements from previous research, as mentioned earlier are considered.

2.5   Evaporative drying

Dewatering is one of the effective procedures for improving the engineering properties of wet soils, and in many cases of embankment construction, evaporation offers considerable possibilities for the economical dewatering of such materials. The results from several evaporative drying studies indicate that evaporation effects are very effective in drying and strengthening wet soil (Willet, 1972; Krizek et al., 1973; Krizek et al., 1978; Benson Jr. and Sill, 1991; Thomas, 1993; Abu-Hejleh and Znidarcic, 1995; and Fujiyasu et al., 2000). The rate of drying depends on many factors such as climatic conditions, initial water content, layer thickness, surface area, and mixing of materials during evaporation.

From the Krizek et al. (1978) study, the relative importance of the various factors that increase the water loss due to evaporation is in the order of 1. mixing, 2. initial water content, 3. weather, 4. exposed surface area, and 5. layer thickness. Several researchers proposed relationships between those factors. For example, U.S. Army Corps of Engineers (1987) provided empirical methods for estimating desiccation behaviour. Benson Jr. and Sill (1991) proposed an approximate closed-form solution to the equation governing the evaporative drying of dredged material. Swarbrick and Fell (1992) described a semi-empirical one-dimensional model of sedimentation and desiccation that includes some restrictive assumptions. Abu-Hejleh and Znidarcic (1995) developed a new desiccation theory and formulated a general form of the finite strain governing equation of the overall consolidation and desiccation process. Seneviratne et al. (1996) described the use of the large-strain consolidation theory as the basis of a computer program for dealing with the problem of consolidation of slurried tailings.

The relationship between water content and shear strength was studied by Krizek et al. (1978) for dredged material drying as shown in Figure 2.4. The study indicated that the relation is linear and quite uniform for water content less than about 60%. However, this study was done for a particular material, thus it seems that the relationship in general is linear but very diffuse and the constant in the relation may vary dependent on the type of soil.

Desiccation behaviour

When very soft clay is placed in a stockpile area volume reduction occurs through two natural processes: self-weight consolidation, and desiccation (evaporative drying). During self-weight consolidation, the upward flux of water through the material is at a fairly rapid rate. While the formation of water on top of the soil material progresses at a rate faster than the evaporation rate, a layer of water will remain over the soil material. When the evaporation rate exceeds the water expulsion rate the soil surface will become exposed. This is the beginning of the evaporative drying phase or desiccation.

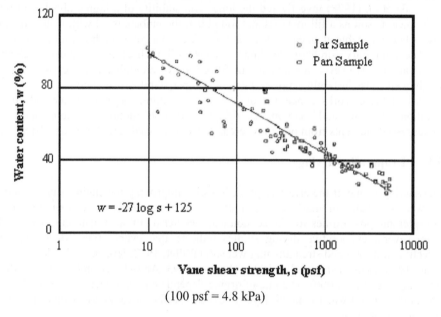

(100 psf = 4.8 kPa)

Figure 2.4: Correlation of shear strength with water content in laboratory evaporation tests (after Krizek et al., 1978)

Desiccation is basically removal of water by evaporation. The potential for evaporation from a saturated surface is controlled by the net radiation at the surface, air temperature, ground temperature, relative humidity, and wind speed. Evaporation from a saturated surface leading to the formation of a desiccated crust may be divided into two phases. The removal of water occurs at differing rates during the two phases as shown in Figure 2.5 and Figure 2.6.

The first phase begins when all free water has been drained from the material surface. The initial void ratio after sedimentation has been empirically determined to be at a water content of approximately 2.5 times the Atterberg liquid limits (w_L) of the material (U.S. Army Corps of Engineers, 1987). The void ratio at this point corresponds to approximate zero effective stress as determined by laboratory sedimentation and consolidation testing on several cohesive sediments. The first phase is called the constant-rate drying period. During the constant-rate drying period, water within the material is transferred to the surface at a rate sufficient to replenish that removed by evaporation. The potential evaporation rate, which is controlled by ambient conditions, will be the rate of drying during the constant-rate drying period. First phase drying ends and the second phase begins at a void ratio that may be called the decantation point or saturation limit (e_{SL}). The e_{SL} has been empirically determined to be at a water content of approximately 1.8 w_L (U.S. Army Corps of Engineers, 1987). This water content is termed the critical moisture content (M_{cr}).

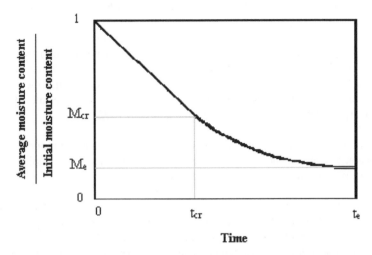

Figure 2.5: Typical moisture content versus time curve

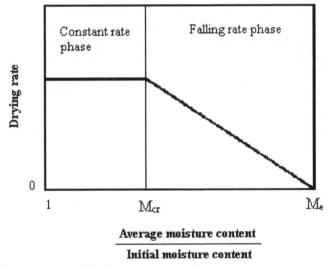

Figure 2.6: Typical drying rate versus moisture content curve

The second-phase drying begins when the material dries sufficiently, the rate at which water is transferred to the surface will not be sufficient to replenish that removed from the surface at the potential evaporation rate. This phase is called falling-rate drying period. During this period, the drying rate continuously decreases and is at a rate determined by conditions within the material, which is drying. The falling-rate drying will be an effective process until the material reaches a void ratio that may be called the desiccation limit (e_{DL}). When the e_{DL} reaches a limiting depth, evaporation of additional water from the material will effectively cease. The e_{DL} may roughly correspond to a water content of 1.2 plastic limit (w_p) (U.S. Army Corps of Engineers, 1987). This moisture content is termed the equilibrium moisture content (M_e).

There are two types of consolidation concerned during desiccation of very soft clay:

*   *Self-weight consolidation*
    Terzaghi theory of one-dimensional consolidation or small strain theory has received widespread application for consolidation problems in which the magnitude of settlement is small in comparison to the thickness of the consolidating layer. However, very soft clay exhibits a very high initial void ratio and very large volume change as loads are imposed. It has been found that the classic Terzaghi theory does not generally apply in these cases. A finite strain theory for one-dimensional consolidation is better suited for describing the large settlements common to the consolidation of very soft clay (Carrier et al., 1983; U.S. Army Corps of Engineers, 1987; Swarbrick and Fell, 1992);

*   *Desiccation consolidation*
    The evaporation of water from the upper parts of the material causes a reduction in its moisture content, which causes a reduction in void ratio or volume occupied due to the negative pore water pressure induced by the drying.

For very soft organic clay in this study, a simulation computer program (RIJPING) (Boels and Oostindië, 1991), for calculation of the water balance, drying, crack formation and surface subsidence of clay sludge is available to determine the relationship among the relevant parameters. These parameters such as thickness, evaporation rate, drying time, moisture content, void ratio, density and shear strength are needed in the optimisation model. RIJPING is a one-dimensional model that calculates the water balance of clay sludge in a vertical profile in time i.e. pressure head, water content, crack volume and layer thickness. For the soil as a whole surface subsidence, total crack volume, groundwater level, drain discharge, matrix infiltration, bypass flow, surface runoff and actual evaporation are computed. The effect of weather conditions (rainfall and evaporation), consolidation fluxes and drainage conditions have been included (Boels and Oostindië, 1991). A schematic representation of the simulation model is shown in Figure 2.7.

In this study, a physical model has been tested on a trial embankment to contribute to the understanding of drying behaviour and evaluate the simulation results associated with the possibility of using very soft organic clay as a fill material for road construction.

2.6   Horizontal sand drainage

The time it may take to consolidate a relatively thick compressible layer can be long. Thus, an accelerated consolidation process by decreasing the drainage path length is one of the methods, which can be used to reduce the time required for settlement to occur. If this method is applied before permanent construction begins, the site can be stabilised and post construction settlement limited to acceptable amounts. Decreasing the drainage path length can be done either vertically or horizontally. For thick strata or natural sub-surfaces, vertical drainage is commonly used, and horizontal drainage is used in embankments.

Many workers (Boman and Broms, 1978; Grace and Green, 1978; Inada et al., 1978; Ostlid, 1981; Lee et al., 1987; and Karunaratne, 1990) studied the potential of using horizontal drainage layers to accelerate the consolidation process. All of the results indicated that embankments with a horizontal drainage layer did show accelerated settlement and efficiently strength gain.

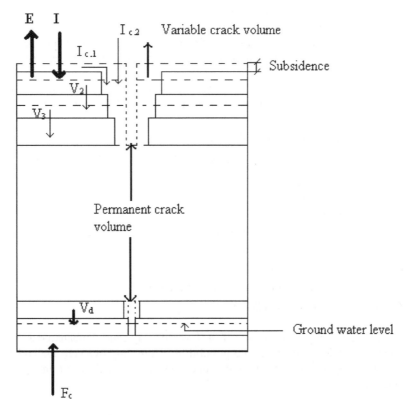

where I   = infiltration rate in soil matrix (cm/d)
       $I_{c,1}$ = part of total infiltration caused by rainfall intensity exceeding maximum infiltration
            rate of soil matrix (cm/d)
       $I_{c,2}$ = part of total infiltration caused by direct entering into cracks (cm/d)
       E   = actual evapotranspiration (m/s)
       V   = Darcy flux between two nodal points (cm/d)
       V_d = flux to groundwater level (drain discharge) (cm/d)
       F_c = consolidation flux (cm/d)

Figure 2.7: Schematic representation of the simulation model RIJPING (Boels and Oostindië, 1991)

Layered clay-sand model

A simplified cross section of a typical layered clay-sand in which a sand layer is sandwiched between two clay layers of equal thickness is shown in Figure 2.8. The relationship of the parameters, which can be involved in the optimisation model, can be determined by a computer simulation program FSCONBAG (Greeuw, 1997). FSCONBAG is a computer program to simulate consolidation of sludge. The consolidation model in this program is based on the so-called Finite Strain Theory (FST), proposed by Gibson (1967) and extended by Gibson (1981). There are several reasons to use this theory instead of the Terzaghi theory. Firstly, in the Terzaghi theory deformation is assumed to be small relative to layer thickness. Obviously this assumption does not hold for very soft clay. In Finite Strain Theory special reduced co-ordinates are employed, thus the deformations are not limited to small values.

Secondly, self-weight is incorporated in Finite Strain Theory, and self-weight often deals with consolidation of very soft soil. Thirdly, the material behaviour is not described by a single consolidation coefficient c_v, but compaction and permeability depend on the void ratio e, which changes during the consolidation process (Greeuw, 1997).

A disadvantage of the Finite Strain Theory is the higher complexity and the non-linearity of the governing equation, which is given below (Gibson et al., 1967):

$$-\frac{\partial e}{\partial t} = \left[\frac{\gamma_s}{\gamma_w} - 1\right] \frac{d}{de}\left[\frac{k(e)}{1+e}\right]\frac{\partial e}{\partial z} + \frac{\partial}{\partial z}\left[\frac{k(e)}{\gamma_w(1+e)} \frac{d\sigma'}{de} \frac{\partial e}{\partial z}\right]$$

(2.4)

where e = void ratio

t = time (day)

γ_s = solid density (kN/m^3)

γ_w = water density (kN/m^3)

k = permeability (m/s)

z = reduced verticle material coordinate (m)

σ' = effective stress (kPa)

A large-scale physical model of a layered clay-sand embankment has been constructed in order to evaluate the possibility of this improvement method for very soft organic clay to contribute to the understanding of behaviour in actural road embankment.

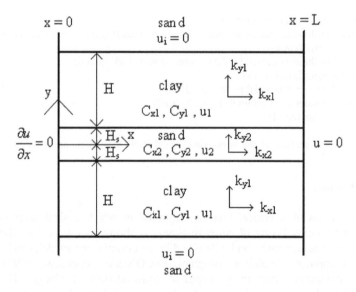

Figure 2.8: Simplified cross section of layered clay-sand

3
Construction equipment and cost analysis

Embankment construction involves movement of a portion of the earth's surface from one location to another and, in its new position, creation of a desired shape and physical condition. Because of the wide variety of soils encountered and work to be done on them, much equipment and considerable costs are involved. This chapter describes and analyses about equipment cost and construction method.

3.1   General equipment for embankment construction

There are many activities involved in embankment construction e.g. excavation, transportation, compaction, and shaping. Equipments that are normally used for these activities especially for working with clay are described. They are bulldozers for grading, hoes for digging and filling, trucks for transportation, and rollers for compaction.

3.1.1   Tractors and tractor accessories

Tractors are designed to provide power for drawbar work. They may be mounted on wheels or crawler tracks. They are essential to many construction projects because they are the prime movers on any construction job where earth must be moved. Typical project applications are land clearing, bulldozing, ripping, and towing other pieces of construction equipment. They are available in a wide range of sizes and power ratings and have an equally wide range of special-purpose attachments. Properly equipped, a tractor usually is the first item moved on to a job and one of the last to finish.

The crawler-(track) type tractor unit is designed for those jobs requiring high traction effort and working on low bearing capacity soil. They are usually rated by size or weight and power. Crawlers are more widely used than wheel tractors (Sain and Quinby, 1996). Table 3.1 provides a comparison of crawler-tractor and wheel-tractor utilisation. An advantage of wheel-type tractors as compared with crawler tractors is the higher speed. However, in order to attain a higher speed, a wheel tractor must sacrifice pulling effort. Also, because of the lower coefficient of traction between rubber tires and some soil surfaces, the wheel tractor may slip its wheels before developing it's rated pulling effort. The crawler tractor is used where high speed is not possible but high drawbar pull and good traction are mandatory. The most essential operating characteristic of the crawler tractor is that it is a powerful, versatile machine best used for rough work over short distances (Russell, 1985). As we consider very soft clay in this study, our attention is drawn to the crawler tractor.

A crawler tractor can be equipped with accessories that enable it to perform a wide variety of tasks. Equipped on the front with a steel blade that can be raised and lowered, the tractor (bulldozer) can push earth from place to place and shape the ground. The blade is used to push, shear, cut, and roll material ahead of the tractor. The straight and angled dozer blades are the prime crawler tractor attachments. They perform a multitude of tasks, from clearing very rough land all the way to finish grading work.

Bulldozers are good machines for stripping and smoothing, which is the removal of a thin layer of material. As with all earth-moving operations, stripping should be conducted in such a manner that push (haul) distances are minimised. Bulldozers are economical machines for haul distances of less than 100 m. The exact economical distance will depend on the material being pushed and on the machine size.

Table 3.1. Tractor-type utilisation comparison (Peurifoy et al., 1996)

Wheel tractor	Crawler tractor
Good on firm soils and concrete and abrasive soils which have no sharp-edged pieces	Can work on a variety of soils. Sharp-edged pieces not destructive to tractor though fine sand will increase running gear wear
Best on level and downhill work	Can work almost any terrain
Wet weather, causing soft and slick surfaces will stop operation	Can work on soft ground and over mud-slick surface; will exert very low ground pressure with special wide tracks and flotation track shoes
The concentrated wheel load will provide compaction and kneading action	
Good for long travel distances	Good for short work distances
Best in loose soils	Can handle tight soils
Has fast return speeds, 13 - 32 km/hr	Slow return speeds, 8 - 11 km/hr
Can only handle moderate blade loads	Can push large blade loads

Equipment manufacturers have developed production formulas for use in estimating the amount of material that bulldozers can push. The following equation is a rule-of-thumb formula proposed by International Harvester (IH) (Peurifoy et al., 1996), which will be used in this study.

$$Production = \frac{net\ hp * 330}{\left(\dfrac{D_b}{0.3048} + 50\right)} * 0.7646 \quad\ldots\ldots\ldots\ldots\ldots\ldots\ldots\ldots(1\ m^3\ per\ hr) \qquad (3.1)$$

where $net\ hp$ = net horsepower at the flywheel for a power-shift crawler tractor
D_b = one-way push distance (m)
$1\ m^3$ = loose cubic meter

To calculate field production rates, one must adjust the value from Eq. 3.1 by the expected job condition. Thus, the equation for calculation of bulldozer production rates is:

Production = Eq.3.1 * product of the correction factors…(1 m³ per hr)   (3.2)

A list of the correction factors is shown in Table 3.2.

3.1.2 Loaders

Loaders are used extensively in construction work to handle bulk material, such as earth and rock, to load trucks, and to excavate earth. There are basically two types of loaders, the crawler-tractor-mounted type and the wheel-tractor-mounted type. They may be further classified by the capacity of the bucket or the weight that the bucket can lift. The wheel loader has a higher production rate than the crawler loader, but can operate only on good conditioned surfaces. If soil is wet, the crawler loader is more suitable. Typical jobs for loaders include loading, unloading, and carrying construction materials.

The production rate for loaders depends on:
*   Fixed time required to load the bucket, shift gears, turn, and dump the load;
*   Time required to travel from the loading to the dumping position;
*   Time required to return to the loading position;
*   Actual volume of material hauled each trip.

Table 3.2. Job condition correction factors for estimating track-type bulldozer production (Last line adapted from Caterpillar, 1995)

Item	Correction factor
Operator	
Excellent	1.00
Average	0.75
Poor	0.60
Material	Material
Loose stockpile	1.20
Very sticky material	0.80
Rock, ripped or blasted	0.60 - 0.80
Slot dozing	1.20
Side-by-side dozing	1.15 - 1.25
Visibility	
Dust, rain, snow, fog, or darkness	0.80
Job efficiency	
50 min/hr	0.83
40 min/hr	0.67
Grade (slope)	$-7*10^{-5}*(\%Grade)^2 - 0.0221*(\%Grade) + 1$

A fixed cycle includes load, manoeuvre, and dump. Fixed cycle times for a track-type loader is 0.25 - 0.35 min and for a wheel loader is 0.45 – 0.70 (Peurifoy, 1996). Haul time and return time can be estimated from the following equations, which will be used in this study:
*   *Track-type loader*

$$Haul\ time = D_l \div \left((0.0144 * hp + 3.69) * \frac{1000}{60} \right) \dots\dots\dots\dots\dots (min) \qquad (3.3)$$

$$Return\ time = D_l \div \left((0.047 * hp + 1.69) * \frac{1000}{60} \right) \dots\dots\dots\dots\dots(min) \qquad (3.4)$$

*   *Wheel loader*

$$Haul\ time = D_l \div \left((0.1214 * hp + 0.27) * \frac{1000}{60} \right) \dots\dots\dots\dots\dots(min) \qquad (3.5)$$

$$\text{Return time} = D_l \div \left((0.1061 * hp + 9.06) * \frac{1000}{60} \right) \dots\dots\dots\dots (min) \qquad (3.6)$$

where D_l = Haul distance (m)
        hp = horsepower at engine flywheel

Total cycle time per load is the sum of fixed time, haul time, and return time. The production rate (1 m^3/hr) for loaders can be estimated as follows:

Production = (60/total cycle time) * bucket fill factor * bucket capacity *
                job efficiency   (3.7)

The bucket fill factor for loaders when the material is soil is 80 - 100% (Peurifoy, 1996). Job efficiency is working time in minutes divided by 60.

3.1.3   Hydraulic excavators

Hydraulic excavators are machines, which make use of hydraulic pressure to penetrate a bucket into the soil. They are classified by the digging motion of the bucket. The hydraulically controlled boom and stick, to which the bucket is attached, may be mounted on either a crawler or a wheel tractor base. A downward arc unit is classified as a "hoe". It develops excavation breakout force by pulling the bucket toward the machine and curling the bucket inward. There is a limitless range of sizes of backhoes, from hoes mounted on small agricultural tractors used in residential construction all the way up to huge crawler-mounted hoes. The work functions of the backhoe often overlap those of other machines such as front-end loaders, tractor-shovels, scrapers, clamshells, and draglines (Russell, 1985). Thus, it is a very versatile machine and commonly used on construction sites.

To estimate backhoe production, we must be able to predict cycle time and average bucket payload. Many factors influence cycle time, including:
*   Size of the machine and bucket;
*   Depth of excavation;
*   Material to be excavated;
*   Dump point (truck, spoil pile, or other);
*   Obstacles on the site;
*   Operator skill.

Some manufacturers produce cycle time charts for various conditions. These charts are general in nature, so that actual job conditions must be taken into account when using such data.

Average bucket payload is fairly easy to come by: manufacturers give bucket specifications, and, once the characteristics of the material are known, bucket capacities may be determined. The following equation may be used to calculate average bucket payload:

Average bucket payload = heaped bucket capacity * bucket fill factor (m^3) (3.8)

Bucket fill factors for some common materials are shown in Table 3.3.
Once cycle time and average bucket payload are determined, the following formula

may be used to estimate backhoe production:

$$\text{Production} = (60/\text{total cycle time}) * \text{average bucket payload}...(1 \text{ m}^3 \text{ per hr})(3.9)$$

Table 3.3. Bucket fill factors for backhoe buckets (Peurifoy et al., 1996)

Material	Bucket fill factor (%)
Moist loam or sandy clay	100 - 110
Sand and gravel	95 - 110
Hard, tough clay	80 - 90
Rock, well blasted	60 - 75
Rock, poorly blasted	40 - 50

This equation represents 100% efficiency (a 60-minute hour); therefore, to estimate actual production the outcome from this equation must be multiplied by the appropriate job efficiency factor. Thus, Eq. 3.5 can be rewritten as:

$$\text{Production} = (60/\text{total cycle time}) * \text{average bucket payload} *$$
$$\text{job efficiency}.................................(1 \text{ m}^3 \text{ per hr}) \quad (3.10)$$

3.1.4 Dump trucks

Dump trucks are the most frequently used equipment for hauling materials over longer distances. They are hauling unit, which, because of their high travel speeds when operating on suitable roads, provide relatively low hauling costs. They provide a high degree of flexibility, as the number in service can usually be increased or decreased easily to permit modifications in the total hauling capacity of a fleet and adjustments for changing haul distances. There are many sizes and types of dump trucks. The type of material to be hauled and the terrain over which it will be transported largely determines the body size and configuration. Dump trucks are available with gasoline or diesel engines with a wide range of power options. They also may have two-wheel or four-wheel drive. Most trucks may be operated over any haul road for which the surface is sufficiently firm and smooth and on which the grades are not excessively steep. Some units now in use are designated as off-highway trucks because their size and total load are larger than permitted on public highways. The capacity of a dump truck may be expressed in at least three ways:
* *Rated capacity* – the load that it will carry, expressed as a weight in tons;
* *Struck volume* – the volumetric amount it will carry, if the load was water level in the body, expressed in cubic meters;
* *Heaped volume* – the volumetric amount it will carry, if the load was heaped on a 2:1 slope above the body, expressed in cubic meters.

The maximum load that can be hauled, in practice, is limited by volume when light loads are carried and by weight when heavy loads are carried.

For the most efficient hauling operation, the size and number of trucks should be balanced with the workload of the excavators and plant processing. That is, in a smoothly run haul cycle, the size and number of trucks should be planned so that the excavators and trucks are kept as uniformly busy as practicable.

The productive capacity of a dump truck depends on the size of its load and the number of trips it can make in an hour. The size of the load can be determined from the specifications furnished by the manufacturer. The number of trips per hour will depend on the weight of the vehicle, the horsepower of the engine, the haul distance, and the condition of the haul road. The productivity can be estimated as follows:

$$\text{Production} = \text{heaped capacity} * ((1/\text{total round-trip time}) *$$
$$\text{job efficiency} \dots\dots\dots\dots\dots\dots\dots\dots\dots\dots\dots (1 \text{ m}^3 \text{ per hr}) \quad (3.11)$$

The time required for each operation in a round-trip cycle is:

$$\text{Loading} = \text{heaped capacity of truck} / \text{loader production} \dots\dots\dots\dots (hr) (3.12)$$

$$\text{Lost time in pit and accelerating} = 0.1 * \text{capacity weight in ton} -1 \dots.. (hr) (3.13)$$

$$\text{Travel to the fill} = \text{haul distance} / \text{haul velocity} \dots\dots\dots\dots\dots\dots (hr) (3.14)$$

$$\text{Dumping, turning and accelerating} = (0.1* \text{capacity weight} - 1) - 0.5.. (hr)(3.15)$$

$$\text{Travel to pit} = \text{haul distance} / \text{return velocity} \dots\dots\dots\dots\dots\dots\dots (hr)(3.16)$$

Haul and return velocity can be determined by the following equation (Karshenas 1989):

$$\text{Velocity} = (\text{hp} * \text{mechanical efficiency} * k) / (\text{gross weight} * \text{total resistance})$$
$$\dots\dots\dots\dots\dots\dots\dots\dots\dots\dots\dots\dots\dots\dots\dots\dots\dots\dots (km \text{ per hr}) \qquad (3.17)$$

where $k = 746$ (N.m/s) (a unit conversion factor)
Gross weight in Newton
Total resistance   = rolling resistance + grade resistance
Rolling resistance   = ((rolling factor * weight in ton)/(weight in ton * 1000))*100
                              (%)
Rolling factor (Peurifoy 1996):
*   25 - 35 kg/m ton for earth, compacted and maintained
*   35 - 50 kg/m ton for earth, poorly maintained
*   75 - 100 kg/m ton for earth, rutted, muddy, not maintained
Grade resistance = haul-road grade (slope) (%)

The production rates of various models of earthmoving equipment that consist of bulldozer, loader, hoe, truck, grader and compactor are shown in Appendix A.

3.1.5 Tamping rollers

A tamping roller may be towed by a tractor or self-propelled, consists of a hollow steel drum on whose outer surface a number of projecting steel feet are welded. The feet on individual rollers may be of varying lengths and cross sections. The weight of a drum may be varied by adding water or sand to produce higher pressure under the feet. As a tamping roller moves over the surface, the feet penetrates the soil to break lumps and produce a kneading action and compact the soil from the bottom to the top of the layer. Compaction of $I_c \geq 0.7$ clay can be done by tamping rollers and produce a high in-situ strength. The production formula for a compactor is:

$$\text{Production} = \frac{W * S * L}{P} * 1.59 \dots\dots\dots\dots\dots\dots\dots\dots\dots\dots(1 \text{ m}^3 \text{ per hr})(3.18)$$

where W = compacted width per roller pass (m)
         S = average roller speed (km/hr)
         L = compacted lift thickness (mm)
         P = number of roller passes to achieve compaction

3.2   Cost analysis

There is a relationship between time to completion of a project and its cost. By understanding the time-cost relationship, the impact of a schedule change on project cost is better to predict. The costs associated with a project can be classified as direct costs and indirect costs. These costs for road embankment construction are described here.

3.2.1   Direct costs

Direct costs include cost of materials, labour, equipment, and supplies required for each component of construction. If the pace of project activities is increased, in order to decrease project completion time, the direct costs generally increase since more resources must be allocated to accelerate the pace. Applying unit prices from supplier quotes to the quantity of materials can determine the cost of materials. In this study, the unit prices of the construction materials are obtained for conditions at hand in the Netherlands. Use is made of a 1999 budget estimate manual, the 'GWW KOSTEN, bemalingen, grondwerken, drainage, 14^e editie' (Riele, 1999). The equipment costs can be determined as follows.

Owning and operating costs

Owning costs involve those costs, which accrue whether or not the equipment is used. Owning costs include interest on capital, which is spent to purchase equipment, the equipment loss value (depreciation), taxes, insurances, and storage expenses.

Operating costs are those costs associated with the operation of equipment. Operating costs usually apply only when the equipment is being used. Operating costs include the fuel and lubrication costs, operator costs, and maintenance and repair costs.

In this study, the Dutch condition was used as reference. Use is made of operating costs derived from "Operating cost standards for construction equipment", 11th revised edition, 1995 (Vereniging Grootbedrijf Bouwnijverheid, 1995) it is used as a reference for owning and operating costs estimation of various equipment types. Equipment types, models and costs that are shown in this book, will be used as equipment data in the optimisation model. The estimation of operating costs is based on assumptions as follows:

*   Standard purchase value (S), the standard rates for depreciation and interest (D + I) and the standard rates for maintenance and repair (M + R) have been determined from technical and statistical data and data relating to business economics valid for the Netherlands in 1995;
*   The estimation does not cover the costs of supervision, storage, insurance, special provisions, and modifications needed in the particular project;
*   Fuel and lubricants price, and operator's wage are at a normal rate for the 1999 level in the Netherlands;

* Standard rates for the costs of depreciation and interest and of maintenance and repair are based on operation of the plant in a single shift per working week;
* Service lift is defined as the period between commissioning of a piece of equipment and the time at which it becomes inefficient to continue to keep the equipment in use;
* In principle, the equipment can be used for a total of 44 weeks per year which implies that 8 weeks is taken as lying idle because of holiday shut-downs. However, a correction has to be made for the time for overhaul or major repair and lying idle awaiting allocation to a job. This implies a degree of utilisation estimated by contractors to be of about 75%, based on 44 weeks availability per year. Thus, the period that the equipment is in use on a site is 32 weeks;
* Equipment has a residual value at the end of its service lift of 5% of the standard purchase value;
* The annuity method is used for calculation of the depreciation. This method is tailored to high investment costs in the light of the usually long lives of the equipment. The interest rate for the purposes of calculation has been taken as 7% annually. An annuity is a constant amount per year, which is made up of depreciation and interest (D and I);
* The weekly costs C for interest and depreciation can be obtained from the following equation:

$$D + I = \left\{ \left(\frac{1}{100} \right) / \left(p^{n} - 1 \right) \right\} \cdot \left(\frac{100}{u} \right) \cdot \left\{ p^{n} \cdot \left(\frac{z}{100} \right) \right\} \ (\% \ of \ S / week) \qquad (3.16)$$

where An   = annuity (%)
     u   = utilisation (weeks/year)
     C   = weekly costs in respect of $D + I$ (€)
     S   = standard value (€)
     n   = service life (years)
     p   = $1 + (i / 100) = 1.07$
     i   = interest rate = 7 (%/year)
     z   = residual value at the end of the service life (% of S)

* Maintenance and repair (M+R) are defined as all activities, which are carried out with the aim of maintaining a system in the technical state necessary for the system to perform properly in respect of the type and extent of its designated functions. The maintenance and repair costs are expressed as a percentage of the standard value. This percentage has been determined based on empirical data. The percentage has been calculated for working in the Netherlands under normal conditions. The costs are derived from interviews with representatives of contractors.

The owning and operating costs of various models of earthmoving equipment that consist of bulldozer, loader, truck, grader and compactor are shown in Appendix B. These costs are not the actual construction equipment costs but are approximate because construction equipment costs must be considered relative to the particular user's operation. Since there can be many variations, it is difficult to generalize about cost computations. However, for a demonstrative study of equipment costs such as this study, these figures are sufficient.

3.2.2 Indirect costs

Indirect costs are overhead costs including transportation of facilities and equipment to the site at the beginning of the job and removing those from the site at the end of the job as well as on site supervision, staffing, operating and maintaining the field office (Stewart, 1991). These indirect costs are an increasing function of the completion time of the project. After all direct costs are calculated and summarised, indirect costs are added to compensate for costs not associated with any single feature of the work and this sum represents the project cost. The typical relationship between construction time, direct costs, and indirect costs is sketched in Figure 3.1.

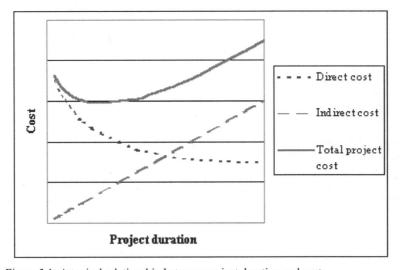

Figure 3.1: A typical relationship between project duration and cost

3.3 Construction method

The construction scheme is shown in Figure 3.2. Clay and sand are stocked in stockpile. The wet clay can be dried in stockpile and agitated by hydraulic excavators before loaded on dump trucks by loaders. The dump trucks carry the clay and the sand to the construction site. The construction of clay-sand layered embankment begins with the placing of the first sand layer on the subgrade. Bulldozers are used to spread the sand into a horizontal layer. The next layer, which is placed on the sand layer, is the clay layer. Bulldozers are also used to spread the clay into a horizontal layer. Compaction by flat or tamping rollers can be done on $I_C \geq 0.7$ clay. After grading and shaping the surface the next sand layer will be placed on top of this layer. The construction method continues in this order until the fill reaches the desired height. In case the clay is very soft and cannot maintain the required side slope, side bunds, which are made from sand or other fill, are constructed alongside the embankment to contain the very soft clay. The typical clay-sand layered embankments are shown in Figure 3.3 and 3.4. Soft clay ($I_C = 0.4 - 0.5$) cannot be compacted by rollers. Instead of mechanical compaction, a surcharge is placed on the embankment after the completion of the embankment to compress the material to a degree that future loadings or repetitions of loadings will not cause further reduction in volume with consequent surface distortions.

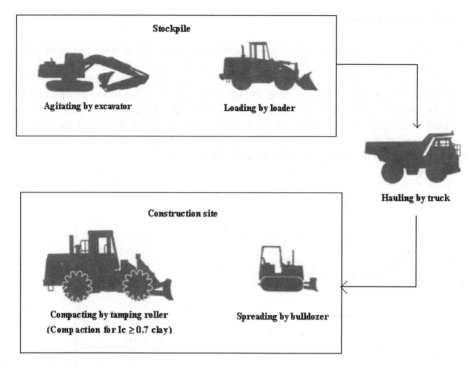

Figure 3.2: Construction scheme

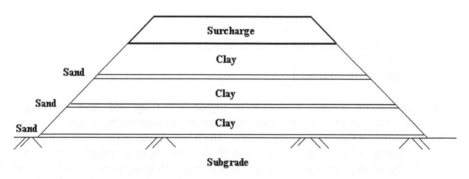

Figure 3.3: Typical clay-sand layered embankment for clay that can maintain a side slope

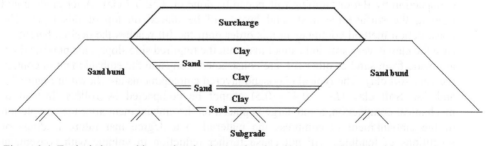

Figure 3.4: Typical clay-sand layered embankment for clay that cannot maintain side slope

4
Optimisation

In a traditional road embankment design process, engineers usually face a wide variety of factors and a huge number of alternatives. These factors such as material properties, dimension of the embankment, traffic load, construction practices, construction time, etc., influence the construction and maintenance costs in some degree and the selection of the factors will determine the economics of the construction. In the traditional design much depends on a trial-and-error determination that requires professional judgment. This procedure is time consuming and the result obtained may not be the economic optimum. Therefore, an approach using an optimisation analysis is introduced in this study in order to reduce the guesswork from the design process and at the same time to yield a cost-optimum solution.

4.1   Optimisation

Optimisation is central to problems involving decision making, whether in engineering or in economics. The task of decision-making entails choosing between various alternatives. This choice is governed by our desire to make the best decision. In a quantitative approach, criteria have to be developed to describe "the best" decision. It is assumed that the measure of goodness of the alternatives can be described by an objective function or performance index. Optimisation theory and methods deal with selecting the best alternative in the sense of the given objective function. Thus, optimisation can be defined as the process of maximising or minimising a desired objective function while satisfying the prevailing constraints (Belegundu and Chandrupatla, 1999). In order to come to the optimal solution a mathematical model for that particular problem can be developed. A mathematical model is the collection of variables and relationships needed to describe pertinent features of such a problem in a mathematical syntax. Optimisation methods then deal with decision problems by formulating and analysing mathematical models. Complex engineering and management problems can use such optimisation methods to gain insight into possible solutions. Engineering optimisation mostly focuses on determining the most cost-effective construction method while satisfying functionality and safety requirements. The choice of a design and construction method can have a significant impact on cost. The objective to minimise construction cost is subject to numerous constraints defining the safe physical limits of the construction materials used, the functionality requirements of the structure and limitation of the equipment and resources (Jacobs, 1997). Although optimisation techniques have been widely used in many areas of civil engineering such as structural, transportation, construction, and environmental, there are limited applications in geotechnical engineering i.e. Chow and Thevendran, 1987, Saribas and Erbatur, 1996, Deutekorn et al., 1997, Abebe and Solomatine, 1998, Berger and Tryon, 1999, Malkawi et al., 2001. However none concerns road embankment design. This study contributes an attempt in using optimisation techniques in road embankment design and construction.

4.2   Mathematical models

Each mathematical model for optimisation usually contains the following elements:
1.   A set of decision variables, which are the design variables that may be controlled

freely by the designer or decision-maker. The set of decision variables must be selected in such a way that all relevant quantities affecting the performance and effectiveness of the system can be evaluated either directly or indirectly. In the mathematical notation used to describe the model the set of decision variables is usually represented by a vector:

$$X = (x_1, x_2, ..., x_n)^T \qquad \text{with x decision variable} \qquad (4.1)$$

where n is the number of decision variables. The object of the analysis is to determine the best possible set of values with respect to system effectiveness. This is termed the optimal policy and is usually denoted by:

$$X^* = (x_1^*, x_2^*, ..., x_n^*)^T \qquad \text{with } x^* \text{ optimum decision variable} \qquad (4.2)$$

2. An objective function, which is the quantity used to measure the effectiveness of a particular policy, is expressed as a function of the decision variables. The aim of the analysis may be either to minimise or maximise the objective function depending on the nature of the problem. However, a problem in minimisation may be converted to one of maximisation by reversing the sign of the objective function. Thus:

$$Minimum[z(X)] = Maximum[-z(X)] \qquad (4.3)$$

3. A set of constraints, which represents the conditions, which must be satisfied before the set of decision variable values, can represent a feasible solution. These feasibility constraints are usually expressed as functions of the decision variables and may be either inequality constraints, e.g.:

$$g_i(X) \geq 0 \quad i = 1, 2, ..., m \qquad (4.4)$$

or equality constraints, e.g.:

$$h_j(X) = 0 \quad j = 1, 2, ..., p \qquad (4.5)$$

The functions $g(X)$ and $h(X)$ may involve any number of the decision variables as well as other quantities and numerical constants. In some classes of problem some or all of the design variables are not permitted to take negative values. This is ensured by including in the statement of the model the required number of non-negativity constraints of the form:

$$x_i \geq 0 \qquad (4.6)$$

Classification of mathematical models

Although the mathematical models for optimisation may generally be described in the format defined above, it is useful to classify certain categories of models in order to better understand the applicability of the various techniques available for their solution. The categories, which are frequently encountered, can be described as follows (Smith et al., 1983; Onwubiko, 2000):

* *Category 1: Constraint*
  If the model is stated with some constraints, we have a constrained optimisation. If, however, the model is stated without constraints, then we have an unconstrained optimisation;
* *Category 2: Linearity*
  If the objective function and all of the constraint functions are linear in terms of the decision variables, then the model is said to be linear. If any of the constraints or any part of the objective function contains a non-linearity the model is said to be non-linear;
* *Category 3: Data*
  This category depends on the nature of the data available. A mathematical model is termed deterministic if all data are assumed to be known with certainty, and probabilistic or stochastic if it involves quantities that are stochastic;
* *Category 4: Variable*
  If the objective function is a function of one variable, we have single-variable optimisation. On the other hand, if the objective function consists of two or more variables, the model is known as multivariable optimisation. If anyone of its decision variables is discrete, implying that the variable is limited to a fixed or countable set of values the mathematical model is termed a discrete programming. If all variables are integers, the model is a pure integer programming; otherwise, it is a mixed-integer programming;
* *Category 5: Time*
  Models, which involve time-dependent interactions, are said to be dynamic, otherwise they are static;
* *Category 6: Objective*
  A mathematical model with a single design objective is known as single-criterion optimisation. But in engineering, it is often a problem to formulate a design in which there are several criteria or design objectives. This model is known as multi-criteria or multi-objective optimisation. If the objectives are opposing, then the problem becomes finding the best possible design, which still satisfies the opposing objectives. An optimum problem must then be solved, with multiple objectives and constraints to be taken into consideration.

Once a mathematical model is considered to represent a system sufficiently, a number of techniques are available to arrive at the optimal solution. In this study complex models have to be made, which are expected to be multi-objective non-linear and mixed integer. Thanedar and Vanderplaats (1995) have stated that available methods for this kind of problem are classified into three categories respectively: branch and bound, approximation using branch and bound, and ad-hoc methods such as simulated annealing and genetic algorithms. The branch and bound method is theoretically correct but is costly to use. Approximation methods provide efficiency but do not guarantee an optimum solution. Ad-hoc methods provide a reasonable solution at an acceptable computational cost but they may be used with the understanding that they usually require considerable user insight and adjustment of algorithm-related parameters to yield reasonable results. To analyse the optimisation models in this study efficiently and obtain global optimisation solutions genetic algorithms (GAs), which is a direct search method, is used as optimisation technique for solving these optimisation problems. GAs are chosen for their ability to search through numbers of local optima and are broadly applicable to optimisation problems that are difficult to solve by conventional techniques. In addition to the discrete and combinatorial nature of the problem, the non-linear behaviour of the cost function and of the soil make optimisation of clay-sand layered embankment complicated. GAs can cope with this particular problem.

There are many factors involved in the construction process of clay-sand layered embankments making the problem very complex. Therefore, the model that is proposed in this study has some simplifications yielding a first approach. GAs also provide a high potential to cope with more complex models. Direct search methods rely solely on evaluation of the objective function at the current position together with experience gained from previous trial positions. All direct search procedures are characterised by the fact that evaluation of derivatives is not involved. The methods are generally robust. A degree of randomness can be introduced in order to achieve global optimisation. Examples of these methods are cyclic co-ordinate search, Hooke and Jeeves pattern search method, Rosenbrock's method, Powell's method of conjugate directions, Nelder and Mead simplex method, simulated annealing (SA), and genetic algorithms (GAs) (Belegundu and Chandrupatla, 1999). In recent years, much research has been devoted to the development and implementation of direct search methods especially GAs, which are being used widely in complex engineering optimisation problems.

4.3   Multi-objective optimisation problems and the Pareto optimality

In engineering problems, one is often confronted with the task of simultaneously minimising or maximising different criteria. When the objectives are opposing, the common difficulty is that none of the feasible solutions reaches an optimum for all objectives simultaneously. Solving these multi-objective optimisation problems concerns finding the best possible solution, which satisfies the objectives and constraints. There are several methods to solve multi-objective optimisation problems. Among these, the ε - constraint or trade-off method, which is one of the most commonly used and simple techniques for solving multi-objective optimisation problems, seems to gain wide acceptance because of its practicality and rationality (Lounis and Cohn, 1993; Ritzel et al., 1994). Hence, this method is used in this study to approach the solution of the problem. This method indicates one of the objectives as the primary objective and constrains the magnitudes of the others. The steps in the solution of a problem are as follows. The general condition is given by

$$Min\ [f_i(\bar{x}),...,f_m(\bar{x})] \qquad\qquad \bar{x}\in\Omega \qquad\qquad\qquad (4.7)$$

where f_i = component objective functions ($i = 1, 2,..., m$)

$\qquad\quad m$ = number of objective functions

$\qquad\quad \bar{x}$ = design variable vector

$\qquad\quad \Omega$ = feasible set to which x belongs

Initially, convert Eq. 4.7 to:

$$Min\ f_1(\bar{x}) \qquad\qquad\qquad\qquad\qquad\qquad\qquad\qquad (4.8)$$

subject to $f_2(\bar{x}) \le \varepsilon_2$ $\qquad\qquad\qquad\qquad\qquad\qquad\qquad (4.9)$

$\qquad\qquad\quad f_m(\bar{x}) \le \varepsilon_m$ $\qquad\qquad\qquad\qquad\qquad\qquad (4.10)$

$\qquad\qquad\quad g_k(\bar{x}) \le 0 \qquad\ (k = 1,..., n_i)$ $\qquad\qquad\qquad (4.11)$

$$h_l(\overline{x}) = 0 \qquad (l = 1, \ldots, n_e) \qquad\qquad (4.12)$$

$$\overline{x} \in \Omega \qquad\qquad (4.13)$$

where ε = assumed value

n_i = number of inequality constraints

n_e = number of equality constraints

Then, the minimum of Eq. 4.8 can be found, subject to an original set and an additional set of constraints. The value of ε is an assumed value, which the designer would prefer not to exceed. Since there is no single optimal solution that simultaneously yields a minimum for all m objective functions, the Pareto optimality is introduced. The Pareto optimality is a solution to the multi-objective optimisation problem. A design variable vector $\overline{x}^* \in \Omega$ is a Pareto optimum for (4.7) if and only if there is no vector $\overline{x} \in \Omega$ with the characteristics $f_i(\overline{x}) \le f_i(\overline{x}^*)$ for all i, $i = 1, 2, \ldots,$ m, and $f_i(x) < f_i(x^*)$ for at least one i, $1 \le i \le m$ (Belegundu and Chandrupatla, 1999). By varying values of ε until a set of acceptable solutions is compiled. It allows the designer to determine the complete Pareto set of optimal points, but only if all possible values of ε are used. The Pareto frontier, which is the set of all of the possible outcomes that are Pareto optimal, can be constructed.

4.4   Genetic algorithms (GAs)

Genetic algorithms (GAs) are different from traditional methods in that they are a probabilistic method that uses a population solution rather than a single solution at a time. Probabilistic is meant here that the algorithms will sometimes accept a bad result in the course of finding an optimum according to a given probability. The beginning of GAs is credited to John Holland, who developed the basic ideas in the late 1960s and early 1970s. He developed computational techniques, which simulated the natural species evolution process and were applied to mathematical programming. Since their conception, GAs have been widely used as a tool in computer programming and artificial intelligence, optimisation, neural network training and many other areas (Chong and Żak, 1996). GAs are robust in that they solve many types of simple as well as complex problems which are of moderate size or are characterised by a mixture of continuous and discrete variables, and their solutions are likely to be a global optimum. GAs are search algorithms that follow the concepts of natural selection and natural genetics (Holland, 1975; Goldberg, 1989). GAs revolve around the genetic operators (reproduction or selection, crossover, and mutation), and "survival of the fittest" strategies.

In general, a GA has five basic components (Michalewicz, 1992):
*   A genetic representation of solutions to the problem;
*   A way to create an initial population of solutions;
*   An evaluation function rating solutions in terms of their fitness;
*   Genetic operators that alter the composition of chromosome;
*   Values for various parameters of genetic algorithms.

A simple GA proceeds by generating an initial population for the first generation at random. The population is considered to consist of a number of chromosomes. A chromosome is made up of a series of characters, which are called genes. This chromosome, which represents a feasible solution to the problem, has to be encoded. Binary encoding is the most common coding method and widely used (Belegundu and Chandrupatla, 1999). For binary encoding, the feasible solution is transformed to a binary string of specific length as shown in Figure 4.1. The length varies with each application.

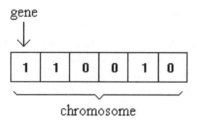

Figure 4.1: Binary encoding

The fitness of each chromosome is evaluated by performing a form of system analysis to compute a value for an objective function. If the solution violates constraints, the value of the objective function is penalized. Some measure of the fitness is applied to construct a better solution in the next generation. Three genetic operations are performed to generate the population of the next generation namely, reproduction (selection), crossover, and mutation.

Reproduction is a process in which chromosomes are duplicated according to their fitness values. The reproduction process is based on the principle of Darwinian natural selection and survival of the fittest. Chromosomes with good fitness value have high probability to survive into the next generation. In this study the so-called roulette wheel selection, which is the best-known selection type, is used. The basic idea of roulette wheel selection is to determine the selection probability for each chromosome proportional to the fitness. In this way, the population of the next generation evolves where the fittest have survived and increase their presence, while the weaker chromosomes die out from the generation.

Crossover is a process in which two members of the newly reproduced chromosomes are selected randomly and exchange part of their chromosome information with a specified probability of crossover (p_c). The exchange proceeds with a selected chromosome, divided into segments at a randomly selected position and one of these segments is exchanged with a corresponding segment of another selected chromosome as shown in Figure 4.2. There are many crossover schemes, such as one-point crossover, multi-point crossover and uniform crossover. There is no theoretical proof as to which one is the best (Pezeshk, 2000). Therefore in this study one-point crossover, which is the simplest one, is used.

Mutation is a process to restore lost or unexplored information into the population to insure that non-existing features from both parent chromosomes may be created and passed on to the next generation. With this operator a larger part of the feasible area of the solution space can be explored. Mutation involves the modification of the content of some genes of a chromosome with a specified probability (probability of mutation, p_m). The mutation process is illustrated in Figure 4.3.

The GA system repeats steps in reproduction, crossover, and mutation in several cycles. The results are feasible solutions that converge stepwise to improved solutions. Figure 4.4 shows the organization of a simple GA.

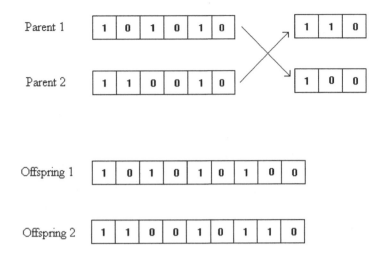

Figure 4.2: Crossover

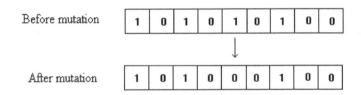

Figure 4.3: Mutation

Penalty Function

The central problem in applying algorithms to constrained optimisation is handling constraints. The search procedure of GAs does not directly consider constraints and genetic operators used to manipulate the chromosomes often yield infeasible offspring. Recently, several techniques have been proposed to handle constraints with GAs. Among these the penalty function technique is most commonly used to handle infeasible solutions in GAs for constrained optimisation problems (Gen and Cheng, 2000). This technique has the advantage that it considers infeasible solutions in genetic search, which appears to be very useful, and can be easily applied to any problem without much change in the algorithm. This constraint management technique allows the search to move through infeasible regions of the search space.

It tends to yield optimisation more rapid and produces better final solutions than do approaches limiting search trajectories only to feasible regions of the search space. The technique transforms the constrained problem into an unconstrained problem by penalizing infeasible solutions, in which a penalty term is added to the objective function for violation of the constraints. The penalty function technique is used to keep a certain amount of infeasible solutions in each generation so as to enforce genetic search towards an optimal solution from both sides of feasible and infeasible regions. The infeasible solutions are simply rejected in each generation, because some may provide useful information about the optimal solution.

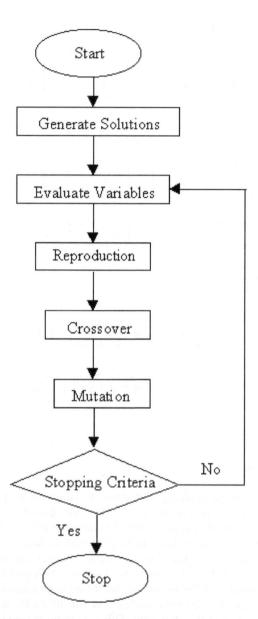

Figure 4.4: Flowchart for a Simple Genetic Algorithm

As mentioned Penalty techniques transform the constrained problem into an unconstrained problem and penalizes infeasible solutions. Although constraints are usually classified as equality or inequality constraints, both types are handled identically in GAs. The GA procedure starts by evaluating the objective function and checks the associated constraints. If no constraint is violated, then no penalty function is assigned to the objective function thus the values of the objective function and its corresponding penalised objective function are identical. In the case that some constraints are violated, a penalty function is applied to the objective function. The value of the penalty is related to the degree in which the constraints are violated. The procedure can be expressed as follows:

$$eval(x) = f(x) + p(x) \qquad\qquad (4.17)$$

where x = chromosome
$\qquad\quad f(x)$ = objective function of problem
$\qquad\quad p(x)$ = penalty term

For minimisation problems, we usually require that:

$$p(x) = 0 \qquad \text{if x is feasible} \qquad\qquad (4.18)$$

$$p(x) > 0 \qquad \text{otherwise} \qquad\qquad (4.19)$$

Several techniques for handling infeasibility have been proposed in the area of GAs. In general, we can classify them into two classes:
*   Constant penalty;
*   Variable penalty.

Variable penalty function approaches are more sophisticated than constant penalty functions. Michalewicz and Schoenauer (1996) concluded that the constant penalty function approach is more robust than the variable penalty function approach. This is because such a sophisticated method may work well on some problems but may not work so well in another problem. In this study a quadratic penalty function, which is the most widely used penalty function, is used to form a penalized objective function. If we consider the following non-linear programming problem:

$$\min \ f(x) \qquad\qquad (4.20)$$

$$\text{Subject to: } g_i(x) \geq 0 \qquad i = 1, 2, \ldots, m \qquad\qquad (4.21)$$

$$eval(x) = f(x) + p(x) \qquad\qquad (4.22)$$

The penalty function is constructed with two components: (1) variable penalty factor and (2) penalty for the violation of constraints as follow:

$$p(x) = 0 \ \ldots\ldots\ldots\ldots\ldots\ldots\text{.if x is feasible} \qquad\qquad (4.23)$$

$$p(x) = \sum_{i=1}^{m} r_i g_i^2(x) \ \ldots\ldots\ldots\text{.otherwise} \qquad\qquad (4.24)$$

where r_i = a variable penalty coefficient for the ith constraint

For each constraint, they create several levels of violation. Depending on the level of violation, r_i varies accordingly. However, determining the level of violation for each constraint and choosing suitable values of r_i is not straightforward and is problem-dependent. An illustration of GAs with penalty functions using a numerical example is given in Appendix C.

4.5   Artificial Neural Networks

Simulation programs i.e. FSCONBAG, MStab, and RIJPING are tools used in calculating aspects of the behaviour of clay and clay embankments in this study. The results of these calculations are an essential part in the optimisation process discussed here. Computation with these simulation programs is time consuming. Furthermore, these programs require much input data on a range of variables. To reduce computational time and input complexity Artificial Neural Networks (ANNs) are introduced. They are used to find patterns in data and relations between input and result. When ANNs are presented with data containing complex patterns they can be trained to identify these patterns. ANNs can fit data where the relation between independent and dependent variables is non-linear and where the specific form of the non-linear relationship is unknown. In this study, the simulation programs FSCONBAG and MStab were run several times with different cases. ANNs models were developed for predicting settlement, settlement time, and stability of the embankment from the simulation programs results. These ANNs models were used in the optimisation process instead of directly linking the simulation programs to the optimisation model.

The principles of ANNs were inspired by a model of the human brain (Russell and Norvig, 1995). In the human brain, the fundamental functional unit, called a neuron, consists of a cell body with dendrites and a long fibre called the axon that has synapses at its end. Neurons are connected to each other through the links of dendrites and synapses. Signals are propagated from neuron to neuron by a complicated electrochemical reaction. A neuron fires a signal through its synapses to the dendrites of the other neurons it is connected to. When the incoming signal raises the electrical potential of a neuron over a certain threshold, this neuron fires.

The ANNs consist of a number of nodes connected by links. Each link has a weight value associated with it. Weights are the means of adapting the ANNs to the problem. Learning in ANNs takes place by updating the weights. The weights are modified in order to tune the ANN behaviour to match the imposed inputs and resulting outputs. Some of the nodes are connected to external points for input and output purposes.

An illustration of a typical ANN unit is given in Figure 4.5. Each unit has a threshold, conveniently represented by a fictitious input to every node. Each node has input links from other nodes and output links to other nodes and each node makes a local computation, based on the cumulative input from its incoming links and outputs a value based on its so-called activation function. The activation function used in this study is the sigmoid function as shown in Figure 4.6 given by:

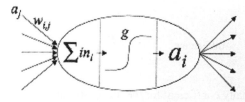

Figure 4.5: A typical ANN unit

$$a_i = g'\left(\sum_i in_i\right) = \frac{1}{1 + e^{-\sum_i in_i}} \tag{4.25}$$

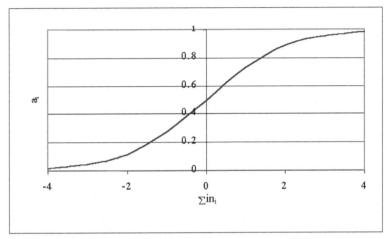

Figure 4.6: Sigmoid function

So-called Multilayer feed-forward ANNs are theoretically capable of representing non-linear functions of any complexity. The structure of such an ANN is illustrated in Figure 4.7.

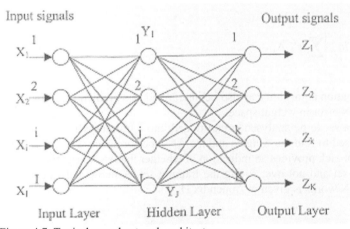

Figure 4.7: Typical neural network architecture

The ANN can be trained by so-called back-propagation learning. Weights are initialised arbitrarily (randomly). Example inputs are presented to the network and if there is an error between the example and the ANN output, the weights are adjusted in such a manner that the error is reduced. The trick in back-propagation learning is to assess the blame for an error and divide it among the contributing weights (Russell and Norvig, 1995). The weight for the link connecting hidden node j to an output node i is updated by:

$$w_{j,i}^{t+1} = w_{j,i}^t + \alpha a_j \Delta_i \tag{4.26}$$

where α is a constant called the learning rate, a_j is the output of node j, and Δ_i is given by:

$$\Delta_i = (T_i - a_i)g'\left(\sum_i in_i\right) \tag{4.27}$$

where T_i = true output,
           a = node output
           g' = derivative of the activation function

Eq. 4.27 can be rewritten as:

$$\Delta_i = (T_i - a_i)a_i(1 - a_i) \tag{4.28}$$

The learning rate is usually close to unity. The learning rate determines the convergence behaviour of the back-propagation algorithm. The optimum value of the learning rate is problem specific. The weight between an input unit k and a hidden layer unit j is updated by:

$$w_{k,j}^{t+1} = w_{k,j}^t + \alpha a_k \Delta_j \tag{4.29}$$

where Δ_j is defined as:

$$\Delta_j = g'\left(\sum_j in_j\right)\sum_i w_{j,i}^t \Delta_i \tag{4.30}$$

The back-propagation algorithm can also be interpreted as a gradient-descent search to minimise the ANN error in weight space.

In practice a data set to be analysed is partitioned into a training set and a test set. The training set is used to find the weights and the ANN performance is tested with the test set. Such an approach provides an indication of whether the ANN is generalising to the whole example set and not over fitting the training set. A numerical example for illustrating how ANN works is given in Appendix D.

5
Optimisation analysis

In this chapter the optimisation process and an example of an optimisation are presented concerning construction of a road embankment constructed with clay and sand as described in previous chapters. The tools enabling the optimisation are presented and discussed as are the parameters used for the case. The chapter concludes with an analysis of the results of the optimisation regarding the various aspects of the construction.

5.1 Outline of the optimisation process

The design variables of a clay-sand layered embankment have been implemented as integer and continuous variables. The relationship of these variables is non-linear. Several mathematical programming methods have been developed for solving the mixed-integer non-linear optimisation problem. However, no single method has been found to be entirely efficient and robust for all different kinds of engineering optimisation problems. With these methods, if there is more than one local optimum in the problem, the result will depend on the choice of the starting point, and a global optimum cannot be guaranteed. Furthermore, when the objective function and constraints are complex functions the search for the solution becomes difficult and is frequently unstable.

A computer program named OLED-GA was developed for **O**ptimising the Clay-sand **L**ayered **E**mbankment **D**esign using **G**enetic **A**lgorithms. This program was coded using Visual Basic 6 language. In this study, the study of optimisation of clay-sand layered embankments was carried out with OLED-GA.

In this chapter, the GA is described that is used to optimise the construction cost and construction time of a clay-sand layered embankment. The design variables consist of clay water content, number of clay layers, clay layer thickness, surcharge thickness, and degree of consolidation. Figure 5.1 shows how the GA process optimises the embankment design. The optimisation process can be described as follows:

* The GA-based design starts by randomly generating an initial population of chromosomes that is composed of candidate solutions to the problem. Each chromosome in the population is a bit string of fixed length;
* After decoding, these chromosomes that represent the design variables are sent to the design part;
* The constraints are checked and if the constraints are violated, the penalty for that step is applied;
* The fitness of each chromosome is evaluated. The new generation is created using the GA operators (reproduction (selection), crossover, and mutation);
* The process is continued until the specified stopping criteria are satisfied.

The program OLED-GA is composed of two main parts: the optimisation part using genetic algorithms and the design part for the clay-sand layered embankment. The structure of the computer program is shown in Figure 5.2 and 5.3. The function of each part is described as follows.

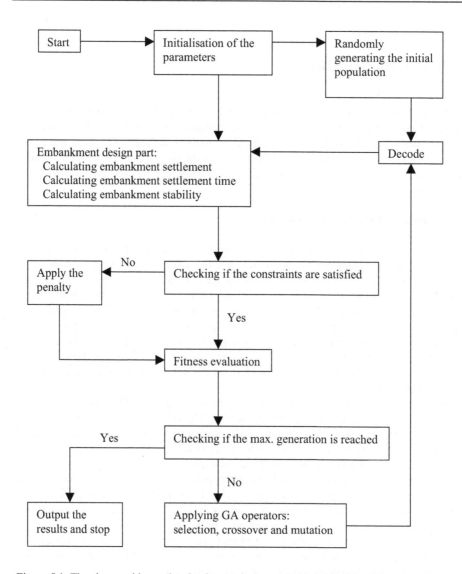

Figure 5.1: The clay-sand layered embankment design optimisation and the GA

The structure of the computer program for optimisation using GA is as follows:

*   *Input module*: with this module, the parameters required by the GA, such as the size of the population, the probability of crossover, the probability of mutation, the generation required to come to a solution are input;
*   *Encode module*: with this module, a number of populations is randomly generated. The binary encoding strategy is used for encoding;
*   *Decode module*: the design variables are calculated from encoded chromosomes and these design variables are sent to the design part;
*   *Design module*: design is evaluated with supplied parameters;
*   *Fitness evaluation module*: the thickness, and the number of clay layers, and the height of surcharge, which are calculated in the design part, are sent to this module. The fitness function, including the effect of the penalty violation is calculated here;
*   *Selection module*: the selection principle based on fitness-proportionate selection is

used in this module to determine the better solution in the next generation;
* *Crossover module*: using one-point crossover schemes, the crossover is carried out in this module. The position of the crossover point is randomly selected. The probability of crossover is provided in the input module;
* *Mutation module*: with a given mutation rate, possible mutation for each of the bits of the chromosomes is performed in this module.

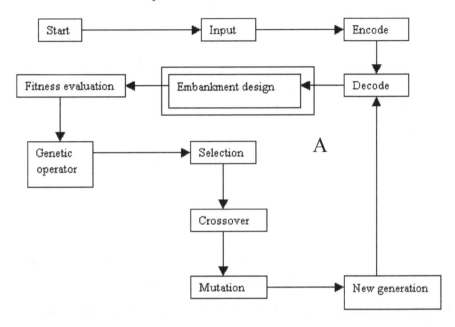

Figure 5.2: Structure of the optimisation part using GA

The structure of the computer program for the design part of the clay-sand layered embankment is as follows:
* *Input module*: the parameters that have to be used for the calculation of the construction cost, the settlement, the settlement time, and the stability of the embankment are input in this module. The parameters for the construction cost calculation are direct cost of clay and sand. The parameters for calculating the settlement, settlement time and stability of the embankment are consistency index, undrained shear strength, initial water content, initial void ratio, and embankment height;
* *Settlement calculation module*: in this module, the settlement of the embankment is calculated. An ANN model for soft clay settlement which was developed from results of the simulation program, FSCONSBAG that will be described in section 5.3, is used for calculating the settlement;
* *Settlement time calculation module*: the time required for the settlement is calculated from an ANN model for soft clay settlement time. The ANN model was developed from the same simulation program FSCONBAG. The detail of this ANN model is described in section 5.3;
* *Stability module*: The stability of the embankment is calculated from an ANN model, which was developed using the results obtained with the slope stability program MStab. Details on the procedure are given in section 5.3.
* *Construction cost module*: in this module, the construction cost of the embankment

is calculated. The output from this module is sent to the fitness calculation module in the GA part.

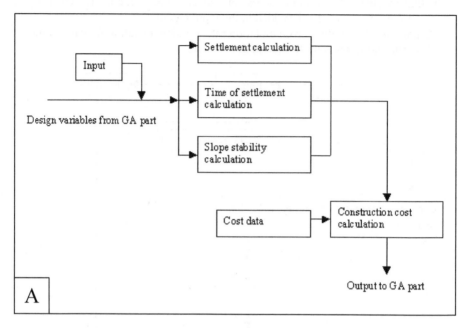

Figure 5.3: Structure of the embankment design part

5.2   The multi-objective formulation of the clay-sand layered construction

The presence of wet clay in earthwork construction may result in considerable problems during the works. Wet clay cannot be compacted to the required dry density. When clay is too wet for compaction, a clay-sand layered construction method can be applied for road embankment construction. The advantage of this method is that horizontal sand layers decrease the drainage path length, which is helpful to decrease the time required for settlement and the related increase in stiffness and strength. To accelerate the consolidation process, and reduce the settlement after completion, a construction temporary surcharge is always used. The studies of Fukazawa and Kurihara (1991), Alonso et al. (2000), and Nash (2001) show that secondary settlement that may occur on organic clay might be reduced by the use of temporary surcharging. Therefore, design of a road embankment constructed using a clay-sand layered scheme with a surcharge, is formulated here.

For a road embankment construction cost and time usually are important. The formulation of the problem as presented here considers minimum construction cost, while the construction time should be minimised. The optimisation of the design of a clay-sand layered embankment involves a number of parameters, which include clay water content, clay layer thickness, sand layer thickness, surcharge thickness, degree of consolidation, service load, basic properties of clay and sand, dimensions of the embankment, haul distance, production rate and cost of materials. Each of these parameters influences the overall construction cost in some degree and the selection of these parameters for the design will determine the economics of the construction. Because of the number of parameters involved, the task of establishing the optimum can be quite complicated. Many options may have to be investigated to ensure a robust solution. The optimisation model, which has been developed and is described here, is intended to provide an optimum solution for the design of a clay-sand layered embankment for road construction, and to permit rapid investigation of the significance of parameters for the optimum design.

The objective of the optimisation is to minimise the direct construction cost, which is defined as the sum of the direct cost functions, while the constraints are satisfied. Using the direct costs as the objective function allows for efficient evaluation of various indirect cost rates without performing another GA run. The direct cost function comprises four components:

*   Cost of clay (at the water content before drying or consolidation);
*   Cost of horizontal sand layers;
*   Cost of surcharge;
*   Cost of sand bunds.

The determination of these cost functions is shown in Appendix J.

The objective function of construction cost is expressed as a function of five design variables:

* w_d = water content of clay after drying (%);
* N_c = number of clay layers;
* C_{in} = initial clay layer thickness (m);
* S = actual surcharge thickness (m);
* D_c = degree of consolidation (%).

The objective function of construction cost can be written as:

$$\text{Min cost} = C_1 + C_2 + C_3 + C_4 \tag{5.1}$$

with the determination of the cost given in Appendix J
For C_1:

C_1 = unit cost of clay * volume of clay

For $I_C \geq 0.7$ clay

$$Unit\ cost\ of\ clay = \left(-6.52 * \left(\frac{w_d}{w_i}\right) + 6.52\right) + (0.68 * d_h + 1.40) \tag{5.2}$$

$$Volume\ of\ clay = (N_c * C_{in}) * (W_r + 2 * (N_c * (C_{in} + S_h))) * L_r \tag{5.3}$$

For $0.5 \leq I_C < 0.7$ clay:

$$Unit\ cost\ of\ clay = \left(-6.52 * \left(\frac{w_d}{w_i}\right) + 6.52\right) + (0.61 * d_h + 2.40) \tag{5.4}$$

$$Volume\ of\ clay = (N_c * C_{in}) * (W_r - 2 * (N_c * (C_{in} + S_h))) * L_r \tag{5.5}$$

For C_2:

C_2 = unit cost of horizontal sand layers * volume of horizontal sand layers

For all clays:

$$Unit\ cost\ of\ horizontal\ sand\ layers = 0.54*d_h + 11.52 \tag{5.6}$$

For $I_C \geq 0.7$ clay:

$$Volume\ of\ sand = (N_s*S_h)*(W_r + 2*(N_s*(C_{in}+S_h)))*L_r \tag{5.7}$$

For $0.5 \leq I_C < 0.7$ clay:

$$Volume\ of\ sand = (N_s*S_h)*(W_r - 2*(N_s*(C_{in}+S_h)))*L_r \tag{5.8}$$

For C_3:

$\quad C_3 \quad$ = unit cost of sand surcharge * volume of sand surcharge

For all clays:

Unit cost of sand surcharge:

$$= (0.54*d_h + 1.52) + 5*(1+r)^{(f(PI,w_dG,C_{in},S,(D_c*total\ settlement)))} - 5 \tag{5.9}$$

Volume of sand surcharge:

$$= S*(W_r - 2*S)*L_r \tag{5.10}$$

For C_4:

$\quad C_4 \quad$ = unit cost of sand bunds* volume of sand bunds

For $0.5 \leq I_C < 0.7$ clay:

Unit cost of sand bunds:

$$= 0.54*d_h + 6.52 \tag{5.11}$$

Volume of sand bunds:

$$= 4 * (N_c * (C_{in} + S_h))^2 * L_r \tag{5.12}$$

where I_c = consistency index
w_d = water content of clay after drying (%)
w_{in} = initial water content of clay before drying (%)
r = rate of return expressed as a decimal
d_h = haul distance (km)
N_c = number of clay layers
C_{in} = initial clay thickness (m)
W_r = road width (m)
S_h = horizontal sand thickness (m)
L_r = road length (m)
N_s = number of horizontal sand layers
PI = plasticity index
G = specify gravity
S = actual surcharge thickness (m)
D_c = degree of consolidation (%)

The objective function of total construction time has five design variables:
* w_d = water content of clay after drying (%);
* N_c = number of clay layers;
* C_{in} = the initial clay thickness (m);
* S = the actual surcharge thickness (m);
* D_c = the degree of consolidation (%).

The objective function for total construction time is:

$$\text{Min time} = t_d + t_c + t_w \tag{5.13}$$

where t_d = drying time
t_c = construction time
t_w = surcharge loading time

as described below

t_d = drying time (day)

$$= -\frac{\left(\dfrac{w_d}{w_i}\right)}{0.0046} + 217.40 \tag{5.14}$$

For $I_C \geq 0.7$ clay:

t_c = construction time of clay layers, horizontal sand layers, and surcharge load

$$= \frac{(N_c * C_{in}) * (W_r + 2 * (N_c * (C_{in} + S_h))) * L_r}{PR_c}$$

$$+ \frac{(N_s * S_h) * (W_r + 2 * (N_c * (C_{in} + S_h))) * L_r}{PR_s} \qquad (5.15)$$

$$+ \frac{S * (W_r - 2 * S)) * L_r}{PR_{sur}}$$

For $0.5 \leq I_C < 0.7$ clay:

t_c= construction time of clay layers, horizontal sand layers, surcharge load and sand bunds

$$= \frac{(N_c * C_{in}) * (W_r - 2 * (N_c * (C_{in} + S_h))) * L_r}{PR_c}$$

$$+ \frac{(N_s * S_h) * (W_r - 2 * (N_c * (C_{in} + S_h))) * L_r}{PR_s} \qquad (5.16)$$

$$+ \frac{S * (W_r - 2 * S) * L_r}{PR_{sur}}$$

$$+ \frac{4 * (N_c * (C_{in} + S_h))^2 * L_r}{PR_{sb}}$$

t_w = waiting time during preloading by surcharge (day)

$$= f(PI, w_dG, C_{in}, S, (D_c * \text{total settlement})) \qquad (5.17)$$

where PR_c = clay production rate (m^3/day)
PR_s = horizontal sand production rate (m^3/day)
PR_{sur}= surcharge load production rate (m^3/day)
PR_{sb} = sand bund production rate (m^3/day)

An ANN model is used to predict the waiting time during surcharging with the parameters as shown in Eq. 5.18. The development of the ANN model for predicting the waiting time during surcharging (settlement time of clay) will be described in section 5.3.

Constraints for these objective functions are considered to be the post construction settlement of the embankment, the final height of the embankment, the surcharge thickness, and the lower-upper limits of the design variables.

The total settlement is the sum of one dimensional consolidation settlement of all clay layers. The settlement occurs due to the weight of clay itself and surcharge load. For the horizontal sand layers no settlement is assumed. The settlement for each layer can be predicted from the ANN model for predicting the settlement of clay, which is described in section 5.3.

$$Settlement_i = f(PI, e_0, C_{in}, L_i) \tag{5.18}$$

where e_0 = initial void ratio = $w_{in}G$ (for saturated condition)
      L_i = applied load on clay layer i (kPa)

In order to eliminate settlement after completion of the construction, the total settlement has to be equal to or to exceed the expected settlement. The expected settlement is to account for the effects of the road pavement structure and the traffic load during service time. The total settlement is calculated from:

$$Total\ settlement = \sum_{i=1}^{N_c} Settlement_i = \sum_{i=1}^{N_c} f(PI, e_0, C_{in}, L_i) \tag{5.19}$$

The applied loads on the top clay layers for the total actual and expected settlement are surcharge and traffic loads during service time respectively. The settlement constraint is:

$$Total\ settlement - expected\ settlement \geq 0 \tag{5.20}$$

and the final embankment height is calculated from:

$$H_f = \sum_{i=1}^{N_c} (C_{in} - settlement_i) + S_h * N_s \tag{5.21}$$

where H_f = final embankment height (m)

and the final embankment height after settlement must be higher than the required height for the road:

$$Final\ embankment\ height - required\ embankment\ height \geq 0 \tag{5.22}$$

The surcharge height constraint can be obtained from the relationship between surcharge height, undrained shear strength of clay, clay layer thickness and embankment height. The prediction of maximum surcharge height can be determined using an ANN model for embankment stability, which is described in section 5.3. and is described as:

$$\text{Maximum surcharge height} = f(C_u, C_{in}, H_e) \tag{5.23}$$

where C_u = undrained shear strength of clay (kPa)
$\quad\quad\ \ H_e$ = embankment height (m)

The surcharge height constraint can be written as:

$$\text{Maximum surcharge height} - S \geq 0 \tag{5.24}$$

The lower-upper limit constraints follow from the description in Chapter 2 are that the clay layer thickness is between 0.30 and 3.8 m, the surcharge height is between 1.5 and 3 m, and the degree of consolidation is between 60 and 90%. All the constraints can be written as follows

$$0.30 \leq C_{in} \leq 3.8 \tag{5.25}$$

$$1.5 \geq S \geq 3 \tag{5.26}$$

$$60 \leq D_c \leq 90 \quad\quad\quad\quad\quad (D_c: \text{degree of consolidation}) \tag{5.27}$$

5.3   The relation between input and output of simulation programs with ANN

Back-propagating neural network models are developed to predict the results from simulation programs on settlement and stability. These ANN models are used to perform the calculations in the design part of the optimisation program. The ANN models are developed and used in the optimisation process in order to reduce computation time of the simulation programs. The ANN models, which are trained using the data from simulation programs, are used to estimate the result from the simulation programs. Three ANN models have been developed for use in OLED-GA, namely: estimating settlement of very soft clay, estimating the settlement time of very soft clay, and embankment stability. The finite strain consolidation simulation program FSCONBAG has been used to develop the models for settlement. The slope stability program MStab has been used to develop the model for slope stability. The development is described in the following section.

5.3.1 Development of the neural network models

The steps for developing ANN models, as outlined by Maier and Dandy (2000), are used as a guide in this study. The steps are the determination of model inputs and outputs, division and pre-processing of the data (see section 5.3.3), the determination of appropriate network architecture, optimisation of the connection weights (training), stopping criteria, and model validation. The personal computer-based software PATHFINDER Version 1.50, by Z Solutions Inc., Atlanta, Georgia, USA, is used for the ANN operation in this study.

The data used to calibrate and validate the neural network models were obtained from FSCONBAG and MStab by performing the calculation with a range of values for the parameters. The ranges for training the ANN are as follows:

* For FSCONBAG, the parameters in the calculation consist of the plasticity index, which is varied from 40 to 80%, initial void ratio is varied from 1 to 3, clay thickness is varied from 0.3 to 3.8 m, and surcharge load is varied from 10 to 200 kPa. An example of FSCONBAG input and output is shown in Appendix E;

* MStab is a computer program, which is developed by GeoDelft (GeoDelft, 2003) to perform slope stability analysis. MStab version 9.7 has been used for a standard embankment slope of 1:2 in this study. The parameters, which are varied, consist of undrained shear strength, clay layer thickness, number of sand layers, and embankment height. The undrained shear strength is varied from 30 to 80 kPa, the clay layer thickness is varied from 0.3 to 3.8 m, the number of sand layers is varied from 1 to 16 layers, and the embankment height is varied from 0.5 to 8 m. The calculation criteria and an example of MStab input and output is given in Appendix F.

5.3.2 Model inputs and outputs

The factors affecting settlement, settlement time, and embankment stability are needed as input in the ANN models. The parameters used for the optimization are chosen as follows:

* Plasticity index, initial void ratio, clay layer thickness, and applied load (surcharge load) are considered to be the factors that affect the settlement, thus they are presented to the ANN model for estimating settlement as model input variables. The settlement value is the output of the model;

* For estimating the settlement time with the ANN model, plasticity index, initial void ratio, clay layer thickness, applied load (surcharge load), and amount of settlement are presented to the ANN model as model input variables. The output from the model is the settlement time;

* For the embankment stability ANN model, undrained shear strength of clay, clay layer thickness, and embankment height are presented as model input variables. The maximum surcharge load, which yields a factor of safety of 2 for a standard slope of 2:1, is the output of the model.

5.3.3 Data division and pre-processing

It is common practice to divide the available data into two subsets; a training set to construct the neural network model, and an independent validation set to estimate model performance in the environment deployed (Twomey and Smith, 1997). However, dividing the data into only two subsets may lead to model over fitting. Thus, the stopping criterion, used to decide where to stop the training process, is used in this study as a tool to ensure that over fitting does not occur (Smith, 1993). The data are randomly divided into three sets: training, testing, and validation. In total, 80% of the data are used for training and 20% are used for validation. The training data are further divided into 70% for the training set and 30% for the testing set (Shahin et al., 2002). The number of data sets (Appendix F) used to calibrate and validate the ANN models is shown in Table 5.1.

Table 5.1. Summary of the data sets used in ANN models development

ANN models	Data sets			
	Training	Testing	Validating	Total
Settlement	158	68	56	282
Time of settlement	608	260	217	1085
Embankment stability	288	124	102	514

It is important to pre-process the data to a suitable form before they are applied to the ANN. Pre-processing the data by scaling them is important to ensure that all variables receive equal attention during training. The output variable has to be scaled to be commensurate with the limits of the transfer functions used in the output layer. In this study, the input and output variables are scaled between $0-1$.

5.3.4 Model Architecture

Determination of the network architecture is important in the development of ANN models. It requires the selection of the number of hidden layers and the number of nodes in each of these. It has been shown that a network with one hidden layer can approximate any continuous function, provided that sufficient connection weights are used (Hornik et al., 1989). Consequently, one hidden layer is used in this study.

The number of nodes in the input and output layers are restricted by the number of model inputs and outputs. Table 5.2 shows the summary of input and output layers of each ANN model. In order to obtain the optimum number of hidden layer nodes for each model, each ANN model is trained with one to ten hidden layer nodes. The smallest network that is able to map the desired relationship is used. The coefficient of determination (r^2), the root-mean-square error (RMSE), and the mean absolute error (MAE) are the criteria used to evaluate the performance of the ANN models. The results of the evaluations are shown in Appendix G. The smallest number of hidden layer nodes, with the highest value of r^2 and the lowest values of RMSE and MEA, is selected as optimum. The evaluation resulted in an optimum number of hidden layer nodes for the ANN models for estimating the settlement, settlement time, and embankment stability are six, eight, and eight nodes, respectively.

Table 5.2. Summary of input and output layers of the ANN models

	Input layer		Hidden layer	Output layer	
ANN model	Num. of nodes	Model inputs	Num. of nodes	Num. of nodes	Model outputs
Settlement	4	Plasticity index (PI)	6	1	Settlement (SET)
		Initial void ratio (eo)			
		Layer thickness (C)			
		Applied load (L)			
Time of settlement	5	Plasticity index (PI)	8	1	Settlement time (TIME)
		Initial void ratio (eo)			
		Layer thickness (C)			
		Applied load (L)			
		Settlement (SET)			
Embankment stability	3	Undrained shear strength (Cu)	8	1	Max. surcharge load (S)
		Layer thickness (C)			
		Embankment height (He)			

5.3.5 Weight optimisation (training)

The process of optimising the connection weights is known as training. A feed forward network and a back-propagation algorithm, commonly used for finding the optimum weight combination, are used in the ANN models. The feed forward networks, trained with the back-propagation algorithm, have already been applied successfully to many geotechnical engineering problems (Goh, 1994, Goh, 1995, Teh et al., 1997, Najjar and Basheer, 1996, Shahin et al., 2002) but were not applied for embankment design.

In this study, the general strategy adopted to find the optimal parameters that control the training process is as follows. For each trial number of hidden layer nodes, random initial weights and biases are generated. The neural network is then trained with momentum terms of 0.2 and learning rates of 0.2 to determine the ANN model.

5.3.6 Connection weights

The neural network program PATHFINDER was run with the criteria, described in the previous sections. The results, the connection weights of the ANN models, are shown in Table 5.3, 5.4, and 5.5.

5.3.7 Model validation

Once the training phase of the model has been accomplished, the performance of the trained model is validated using the validation data, which have not been used as part of the model building process. The purpose of the model validation phase is to ensure that the model has the ability to generalize, within the limits set by the training data, rather than simply having memorized the input-output relationships that are contained in the training data.

Table 5.3. The connection weight of the ANN model for settlement

Input/Hidden

	PI	e_0	C	L	Bias
Hidden - 1	-0.227	2.270	-2.999	0.515	-1.669
Hidden - 2	-0.316	-0.022	-0.753	-0.501	-0.340
Hidden - 3	0.117	1.162	-0.755	0.288	0.216
Hidden - 4	0.577	-2.790	-1.164	-0.829	1.933
Hidden - 5	0.584	-1.893	-2.045	-0.991	4.874
Hidden -6	-0.232	-0.235	-0.693	-0.481	-0.424

Hidden/Output

	Hidden - 1	Hidden - 2	Hidden - 3	Hidden - 4	Hidden - 5	Hidden - 6	Bias
SET	-2.326	0.740	1.982	-1.365	-3.268	0.445	0.445

Table 5.4. The connection weight of the ANN model for settlement time

Input/Hidden

	PI	e_0	C	S	SET	Bias
Hidden - 1	-1.389	1.320	0.099	1.144	-4.312	-0.021
Hidden - 2	0.700	-1.320	-2.161	-0.509	-3.887	0.160
Hidden - 3	-0.313	-0.709	-1.091	-0.303	-1.973	-0.474
Hidden - 4	-2.784	2.136	2.942	1.007	-4.415	-0.566
Hidden - 5	-2.719	5.829	1.625	1.772	-2.441	-1.183
Hidden -6	-1.972	1.749	1.428	0.575	-2.707	-0.948
Hidden –7	0.962	0.298	-1.523	0.427	-5.866	0.243
Hidden –8	1.664	-2.172	-1.752	-1.039	-5.765	-0.352

Hidden/Output

	Hid - 1	Hid - 2	Hid - 3	Hid - 4	Hid - 5	Hid - 6	Hid- 7	Hid - 8	Bias
TIME	-2.091	-2.652	-0.917	-2.452	-2.723	-1.354	-3.518	-4.399	3.787

Table 5.5. The connection weight of the ANN model for embankment stability

Input/Hidden

	C_u	C	H_e	Bias
Hidden - 1	-1.191	4.114	0.007	2.463
Hidden - 2	-0.330	-0.723	-1.010	0.913
Hidden - 3	-0.715	-0.103	-0.245	-0.417
Hidden - 4	0.040	-0.018	-0.127	-0.985
Hidden - 5	-0.195	-0.454	-0.578	-0.040
Hidden -6	2.490	0.953	-1.121	-3.556
Hidden –7	-4.188	0.259	4.344	-3.319
Hidden –8	-1.438	1.231	0.183	0.225

Hidden/Output

	Hid - 1	Hid - 2	Hid - 3	Hid - 4	Hid - 5	Hid - 6	Hid- 7	Hid - 8	Bias
S	2.874	-1.311	-0.718	-0.026	-0.851	2.447	-1.817	-1.690	-0.635

The coefficient of determination (r^2), the root-mean-square error (RMSE), and the mean absolute error (MAE) are used to evaluate the performance of the ANN models.
*   The ANN model for settlement has $r^2 = 0.9881$, RMSE = 0.021 m, and MAE = 0.05 m. Comparison between validation data and predicted data from the ANN model is shown in Figure 5.4. This outcome is sufficiently accurate for the optimisation study;

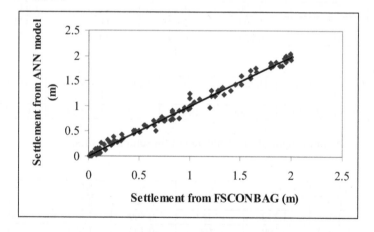

Figure 5.4: Comparison of settlement from FSCONBAG and the ANN settlement model

*   The ANN model for settlement time has $r^2 = 0.9642$, RMSE = 0.0187 days, and MAE = 18.07 days. Comparison between validation data and predicted data from the ANN model is shown in Figure 5.5 which is sufficiently accurate for the optimisation study;

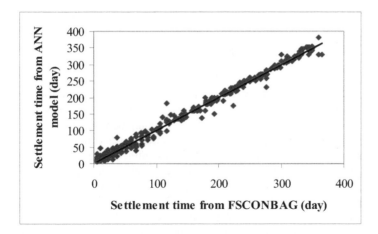

Figure 5.5: Comparison of settlement time from FSCONBAG and ANN settlement time model

*   The ANN model for embankment stability has $r^2 = 0.9987$, RMSE = 0.0064 kPa, and MAE = 0.93 kPa. Comparison between validation data and predicted data from the ANN model is shown in Figure 5.6 which is sufficiently accurate for the optimisation study;

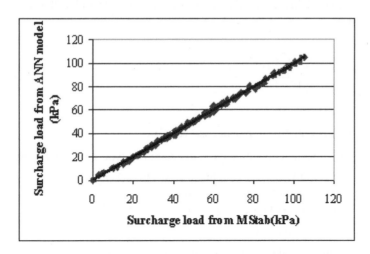

Figure 5.6: Comparison of surcharge load determination from MStab and ANN stability model

It can be seen that the ANN models have the ability to get the results closely to the results calculated by simulation programs. By using ANN models, computational time can be highly reduced making it possible to schematise the non-linear soil behaviour in the optimisation scheme.

5.4   Direct costs optimisation

The Direct costs for a clay-sand layered embankment consists of material costs, equipment costs and labour costs. The materials are clay and sand; estimates of their costs have been obtained from GWW KOSTEN (1999). The equipment fleet is considered to consist of bulldozers, loaders, excavators, dump trucks and compactors. At the stockpile, excavators are used to agitate the clay during the drying period. Loaders or excavators are used to lift the material to dump trucks. Dump trucks carry the material to the embankment area. Bulldozers are used for spreading, grading, and shaving the layers. For the clay which has the $I_C \geq 0.7$ compaction has to be done by rollers or compactors in order to break clay lumps and decrease air voids to obtain sufficiently strong and stiff base. A careful selection of equipment size and number can result in substantial effect both in time and cost. The direct costs can be reduced if the proper equipment size and number are selected. Many contractors select the equipment and estimate the production rate from their experience (Amirkhanian and Baker, 1992). Thus, their selections heavily depend on skilled judgement, but which may not be an optimal selection. The optimal solution or the proper size and number of the equipment fleet for earth-moving operation can be determined by GAs (Limsiri et al., 2003). In this study this approach is used to determine the optimum direct cost for the equipment fleet and this cost can be used in the optimisation analysis of the multi-objective optimisation model of clay-sand layered embankment as discussed in the next paragraph 5.4.1.

5.4.1 Optimisation formulation

The objective of the problem is to select the size and composition of an earthmoving equipment fleet, which yields the minimum equipment costs. The objective function can be expressed as:

$$Minimum\ C_{eq} = \frac{\sum_{i=1}^{n} \sum_{j=1}^{m} (C_o + C_w)_{ij} * \alpha_{ij} N_{ij}}{PR} \qquad (5.28)$$

where C_{eq} = construction equipment cost (€/m³)

C_o = owning and operation costs of equipment type i model j (€/hr)

C_w = operator wages for equipment type i model j (€/hr)

PR = required production rate (m³/hr)

α_{ij} = discrete variable: selected or not selected equipment type i

model j (0 or 1)

N_{ij} = number of the equipment type i model j used

m = number of the equipment model, n = number of the equipment type

Constraints of this problem are production rate, number of equipment, and model of the equipment. The constraints can be expressed as:

$$\sum_{j=1}^{m} PR_j * \alpha_j N_j \geq PR \qquad \text{(for each equipment type)} \qquad (5.29)$$

$$\sum_{j=1}^{m} N_j \leq \text{Maximum number of each equipment type} \qquad (5.30)$$

$$N_j > 0 \text{ and integer} \qquad (j = 1, \ldots, m) \qquad (5.31)$$

$$\sum_{j=1}^{m} \alpha_j \leq \text{Maximum model of each equipment type} \qquad (5.32)$$

$$\alpha_j = 0 \qquad \text{(not selected when } j = 1, \ldots, m) \qquad (5.33)$$

$$\alpha_j = 1 \qquad \text{(selected when } j = 1, \ldots, m) \qquad (5.34)$$

where $\quad PR_j \quad$ = production rate of equipment model j (m^3/hr)

It can be remarked that operation costs can be reduced if a larger fleet operates. This variation is not incorporated in the scope of this thesis.

All related items of the optimisation model are formed into a conceptual model as shown in Figure 5.7.

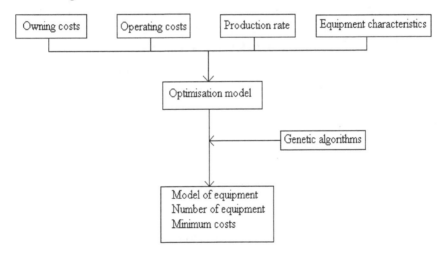

Figure 5.7: Conceptual model for equipment costs optimisation

5.4.2 Optimisation results of equipment selection

Design variables of the model for loaders, trucks, bulldozers and compactors are the number of equipment of each model (N_{ij}) and discrete values for selecting or not selecting (α_{ij}) the model and type. Parameters of the optimisation model are owning and operating costs (C_o), operator wage (C_w), production rate (PR_j) of each model of loaders, trucks, bulldozers and compactors which are listed in Appendix A and Appendix B, and required production rate (PR) of the project. The GAs as described in section 4.4 was applied to the optimisation model described in section 5.4.1 for the selection of the equipment of the clay-sand layered embankment construction with criteria as follows. Clay with $I_C > 0.7$ is the material that has to be moved from stockpile to dumping site 2 km away. The road for truck hauling is an earth road with poor maintenance and no grade. Working condition is average with job efficiency of 0.83. Operator wage is 20 €/hr for each equipment type. The generated initial population of the solutions consists of 360-members. One chromosome consists of 144-bit string as follows. The number of loaders, bulldozers and compactors for each model (N_{ij}) is represented by a 3-bit string. A 5-bit string represents the number of trucks for each model. The selected or not selected parameter for the equipment model j type i is represented by a 1-bit string. The probability of crossover = 0.8; probability of mutation = 0.007; number of generation = 400. The selection of the GA parameters is described in Appendix H.

By varying the required production rate of the project between 50 to 1000 m³/hr with 50 m³/hr intervals, the set of optimal solutions, which is obtained using the GAs, is plotted in Figure 5.8. Table 5.6 shows some samples of the optimum result from the GA.

In general the minimum equipment cost occur when the production rate is higher than 200 m³/hr. A small number of large size equipment is selected rather than a large number of smaller sized equipment for the same production rate. Figure 5.8 shows that a production rate of more than 200 m³/hr should be selected for this project. The optimum results for $I_C < 0.7$ clay and sand can be seen in Figure 5.9 and 5.10.

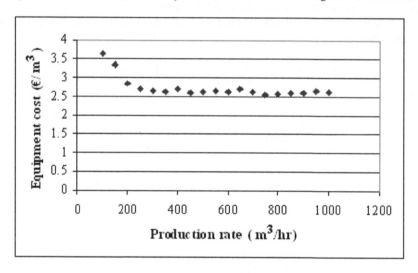

Figure 5.8: Optimal solutions of equipment selection for working with $I_C > 0.7$ clay

The result of this study indicates that this optimisation model and GAs can find an optimum cost and composition of fleet of equipment for a specific production rate. The results appear realistic. This approach is applied to determine the direct cost function, which is one of the parameters that have been used in the clay-sand optimisation model. The determination of the direct cost function is given in Appendix J.

Table 5.6. Optimum results of equipment selection for $I_C > 0.7$ clay

Production rate	100	500	800

(m^3/hr)			
Wheel loader (number-hp)	1-70	2-160	2-215
Truck (number-hp)	3-220	3-220 and 7-280	2-220 and 11-280
Bulldozer (number-hp)	1-180	1-135 and 2-280	1-135 and 5-280
Compactor (number-hp)	1-77	1-107 and 1-145	1-77 and 2-145
Equipment cost (€/m^3)	3.62	2.62	2.60

5.5   Solution approach of clay-sand layered embankment model

Since costs are considered to be the most significant factor for making decisions in most construction projects, the parameters determining the cost are considered suitable to cover all activities in construction. Therefore, the ε-constraint (see par.4.3) approach with the minimum cost as the primary objective is applied here. The multi-objective optimisation model for a clay-sand layered embankment, which was described in section 5.2, can be transformed into a single objective optimisation model by changing the minimum construction time into a constraint as follows:

$$\text{Min cost} = C_1 + C_2 + C_3 + C_4 \tag{5.36}$$

Subject to:

$$\text{Min time} \leq \varepsilon \tag{5.37}$$

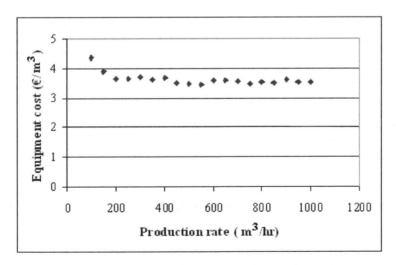

Figure 5.9: Optimal solutions of equipment selection for working with $0.5 < I_C < 0.7$ clay

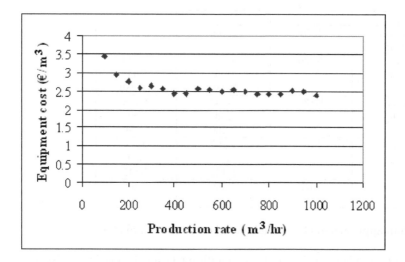

Figure 5.10: Optimal solutions of equipment selection for sand

The set of Pareto optima is obtained by varying values within the constrained range. The bounds on ε are determined from the allowed construction time for a road embankment construction. The lower bound is specified as 90 days, and 360 days is used for upper bound in this study.

The computer program, OLED-GA was used to perform the optimisation process. The program was developed for the design of the two types of clay-sand layered embankments, which were described in section 3.3. Design variables of the optimisation model are the water content of clay after drying (w_d), the number of clay layers (N_c), the initial clay layer thickness (C_{in}), the actual surcharge thickness (S), and the degree of consolidation (D_c). Parameters of the optimisation model are embankment height, clay properties (initial water content, unit weight, specific gravity, liquid limit, plastic limit, plasticity index), sand property (thickness, unit weight), loads during service time (pavement structure load and traffic load), haul distance, production rate, direct cost (direct cost of clay, direct cost of horizontal sand, direct cost of sand bunds, direct cost of surcharge load). The GA as described in section 4.4 was applied on the optimisation model in section 5.2 with a criterion as follows. An initial population of solutions of 84-members is used. One chromosome consists of a 33-bit string. A 6-bit string represents the water content after drying. A 4-bit string represents the number of clay layers. A 9-bit string represents the thickness of clay layers. An 8-bit string represents the surcharge thickness. A 6-bit string represents the degree of consolidation. Probability of crossover = 0.8; probability of mutation = 0.03; number of generation = 100. The selection of the GA parameters is described in Appendix H.

5.6   Optimisation results of clay-sand layered embankment

This section presents the results of the investigation on the optimum design of a clay-sand layered embankment. The optimum design problem was solved by the program "OLED-GA", developed in this study. Two objective functions, namely, the construction cost and the construction time of the embankment were considered in the multi-objective optimisation function. The formulation and algorithm were tested on several problems, which are discussed later in relation to a sensitivity analysis of the design.

A representative case is presented here. The parameters of the example problem are tabulated in Table 5. 7. The clay parameters were chosen after laboratory test results of very soft organic clay taken from a dredging sludge disposal pond, the Slufter Rotterdam. Details of the laboratory test results are given in section 6.1. The cost parameters were obtained from the average equipment cost as described in section 5.4.2 plus material costs obtained from GWW KOSTEN (1999). The GA-based optimisation program "OLED-GA" was run 5 times and the resulting optimum solutions, objective function values, magnitude of design variables, and magnitude of constraints, are shown in Table 5.8, 5.9 and 5.10 respectively. For each run, the GA started from a different initial population in order to avoid local optima.

The optimum result is € 1,752,895 for the construction cost and 164 days for the construction time when the required construction time is limited to 180 days. It can be seen from the results that the formulation and algorithm give a robust solution and then an optimal or nearly optimum point can be found in this complex solution space. The set of optimal solutions, obtained from OLED-GA for the construction time 120 – 180 days, is plotted in Figure 5.11 showing the optimal Pareto frontier. The set of the optimal solutions of the design variables is shown in Figure 5.12 - 5.16.

The Pareto frontier, Figure 5.11, shows that the relationship of the two objectives is not linear. The optimum solutions can be divided into two sections respectively with a construction time between 120 and 150 days and between 150 and 180 days. The Pareto frontier shows that a short construction time involves higher construction cost than a long construction time.

Table 5.7. Input parameters for the example problem

Input parameter	Unit	Value
Required construction time	day	180
Required final embankment height	m	4
Initial clay water content	%	100
Liquid limit of clay	%	100
Plastic limit of clay	%	40
Specific gravity of clay	-	2.6
Unit weight of clay	kN/m^3	15
Sand layer thickness	m	0.20
Unit weight of sand	kN/m^3	18.5
Service load	kPa	35
Haul distance	km	2
Road width	m	25
Road length	m	2000
Production rate for clay layers	m^3/day	4000
Production rate for horizontal sand layers	m^3/day	6000
Production rate for sand bunds	m^3/day	6000

Table 5.8. Optimum values of the objective function for the example

Run-number	Construction cost (€)	Construction time (day)
1	1,752,895	164
2	1,752,895	164
3	1,844,828	162
4	1,752,895	164
5	1,844,828	162

Table 5.9. Optimum values of design variables for the example

Design variables	Unit	Lower bounds	Upper bounds	Optimum values				
				Run 1	Run 2	Run 3	Run 4	Run 5
Water content after drying	%	40	70	55	55	56	55	56
Number of clay layers		2	14	2	2	3	2	3
Clay thickness	m	0.30	3.8	1.87	1.87	1.18	1.87	1.18
Surcharge thickness	m	1.5	3	0	0	0	0	0
Degree of consolidation	%	60	90	60	60	60	60	60

Table 5.10. Value of the constraints at optimum values of design variables for the example

Run		Allow settlement (m)	Final height (m)	Surcharge thickness (m)
1	Limit	≤ 0.20	≥ 4	≤ 2.62
	Obtained value	0.19	4	0
2	Limit	≤ 0.20	≥ 4	≤ 2.62
	Obtained value	0.19	4	0
3	Limit	≤ 0.20	≥ 4	≤ 2.49
	Obtained value	0.20	4.02	0
4	Limit	≤ 0.20	≥ 4	≤ 2.62
	Obtained value	0.19	4	0
5	Limit	≤ 0.20	≥ 4	≤ 2.49
	Obtained value	0.20	4.02	0

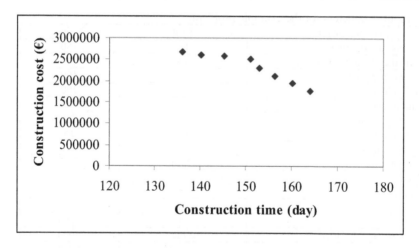

Figure 5.11: Pareto frontier of construction cost and construction time

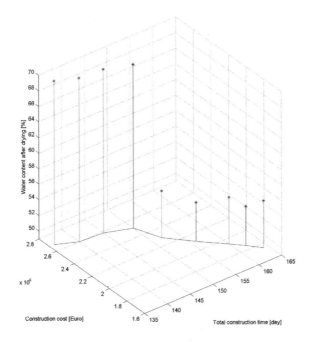

Figure 5.12: Optimum solutions of water content after drying

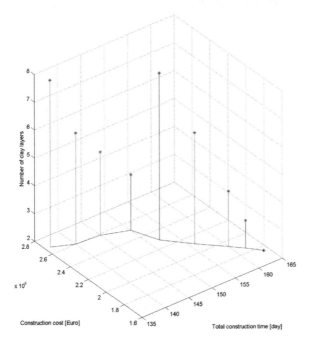

Figure 5.13: Optimum solutions of number of clay layers

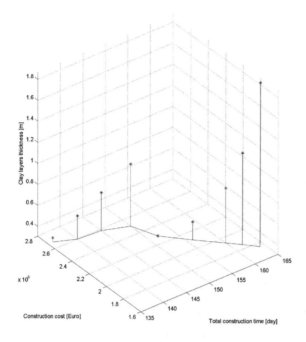

Figure 5.14: Optimum solutions of clay layers thickness

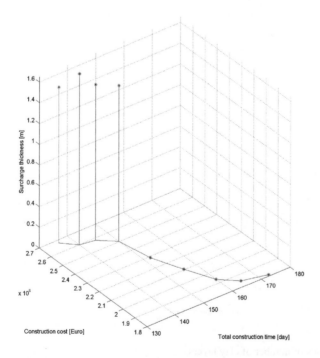

Figure 5.15: Optimum solutions of surcharge thickness

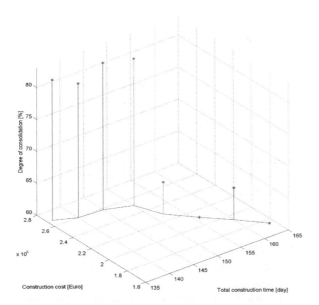

Figure 5.16: Optimum solutions of degree of consolidation

Due to different type of embankment is used, there are two discontinuous sections can be distinguished in the graphs (Figure 5.12-5.16). It can be seen from the optimum solutions of design variables that the first section represents the optimum solutions of the embankments using a high water content clay ($I_C = 0.5$) for construction (embankment with sand bunds). This clay is used when a short construction time is required. A high water content, after drying, and using of many horizontal sand layers appears the preferable choice. For wet clays the required total construction time increases as the number of sand layers decreases and consequently the construction cost reduced. The second section represents the optimum solutions of the lower water content clay ($I_C \approx 0.75$) (embankment without sand bunds). This clay requires construction times longer than 150 days, this includes the period for drying of the sludge to this water content, and construction period. The construction cost is also reduced if the required construction time is increased. There are two sections in the Pareto frontier because of the change in construction method from an embankment with sand bunds in the case of $I_C < 0.7$ to an embankment without sand bunds in the case of $I_C > 0.7$. For both types of embankment reduction in sand layers brings a reduction of construction cost but, consequently, the construction time is longer. Surcharge load which is not necessary for clay of $I_C \geq 0.75$ that induce a small settlement also reduces the construction cost of embankment without sand bunds. The sand layers in the dry clay are redundant for constructing the embankment with such clay, but which may serve as safety measure in case water is able to penetrate the construction during its lifetime. Covering the embankment with a suitable thickness of cover material such as sand, makes the slopes stand in time, if the cover is omitted, the slopes will suffer shallow failures in time (10 - 15 years after construction) as is often seen in steeper clay slopes and which is due to penetration of weathering and soil formation in the slope.

For the embankment with sand bunds, the thickness of clay layers is between 0.35-0.90 m., and for the embankment without sand bunds the thickness is between 0.31-1.87 m. Thus, it seems that the suitable clay layer thickness should not be more than

0.90 m for the embankment with sand bunds and can be thicker for the embankment without sand bunds in this case. To achieve all constraints within a specific construction time the embankment has to be constructed with layers as shown in Figures 5.13 and 5.14. If the number of clay layers and clay layer thickness are chosen incorrectly the requirements of the design will not be met. Figures 15.15 and 15.16 show the surcharge thickness with the degree of consolidation to be used to achieve minimum construction cost. The surcharge load is needed for the embankment with sand bunds to induce the embankment settlement equal to the requirement. For embankments without sand bunds due to a small settlement because of using dry clay the surcharge load is not necessary.

The optimum solution shows that at a specific construction time the number of clay layers, clay layer thickness, and corresponding surcharge thickness has to be chosen properly in order to yield minimum construction cost.

5.7   Sensitivity analysis

The approach followed allows for analysis of the sensitivity of the optimum design. The sensitivity of the optimum solution with respect to a variety of design parameters may be searched. Sensitivity of the optimum solution to changes in these parameters is an important issue for practical design. In this study the sensitivity of the optimum solution with respect to dimension of the embankment, material property and unit cost of materials has been analysed. The results of the analyses are discussed, including the sensitivities of the optimum construction cost and optimum construction time as objective functions and the optimum values of the design variables. The sensitivities of the objective functions are explained for the design parameters considered.

The variation of the parameters considered in the sensitivity analysis is given in Table 5.11. The representative embankment of section 5.6 is used for the sensitivity analysis by varying parameters for each item shown in the table. Those varying parameters are used as the input for the optimisation model in OLED-GA. The OLED-GA was run 5 times for each case and the best solution was selected. The significant effects on the objective functions and the optimum values of design variables with respect to change in parameters mentioned above are given and discussed in the following section.

Table 5.11. Input parameters for sensitivity analysis

Input parameter	Unit	Value
Embankment final height and corresponding road length (maintaining constant volume of the embankment)	m-m	3-2840, 5-1510
Plasticity index	%	50, 70
Unit cost of clay layer	€/m^3	2 times of normal cost
Unit cost of sand layer	€/m^3	2 times of normal cost

5.7.1   Sensitivity on dimension of the embankment

The results of the sensitivity on dimension of the embankment presented in Figure 5.17 which show that for all embankment heights two types of embankment were used. The embankment with sand bunds is selected for short construction time periods that require construction time less than 150 day. At a same construction time the higher embankments can be constructed with relatively lower construction cost than lower embankments with the same volume and greater lengths. The higher costs come from the horizontal sand layer cost and surcharging cost of the lower embankments, which is higher than for the higher embankments at the same specific construction time. These costs have more influence on the total construction cost than cost of clay as can be seen in Figure 5.18 where the construction cost of embankment parts have been plotted for various heights at a construction time of 140 days. The type of embankment was changed from embankment with sand bunds to embankment without sand bunds when the construction time is more than 150 day. The construction time and the construction cost of higher embankments become relatively higher than lower embankments. This comes from the cost of horizontal sand layer as shown in Figure 5.19. At the same specific construction time all embankments used nearly the same clay layer thickness and a higher embankment has more horizontal sand layer than a lower embankment as shown in Figures 5.20 and 5.21.

The height variation does not show any effect with respect to the water content after drying, surcharge thickness, and degree of consolidation. Figures 5.22 and 5.23 show that the water content of 67-70% (I_C = 0.55-0.50) is the optimum solution for all embankments with sand bunds and the water content of 53-56% (I_C = 0.78-0.75) is the solution for all embankments without sand bunds. For each type of embankment only little variation of the water content can be noticed for all heights. The optimum results of surcharge thickness are shown in Figures 5.24 and 5.25. The variation of height has no influence on the surcharge thickness either. The similar surcharge thickness at a specific construction time for all heights of embankment with sand bunds results in a higher surcharge volume on comparison to the lower embankments which causes the lower embankment to have higher construction cost.

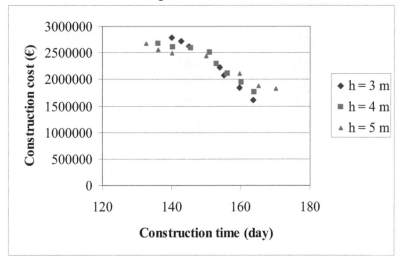

Figure 5.17: Effect of embankment height for construction cost

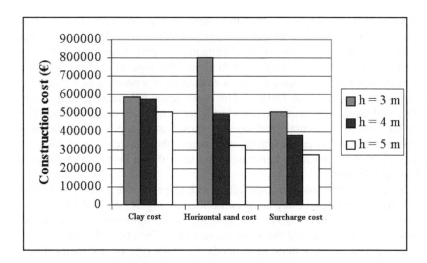

Figure 5.18: Effect of embankment height on the construction cost of embankment with sand bunds parts at 140 days

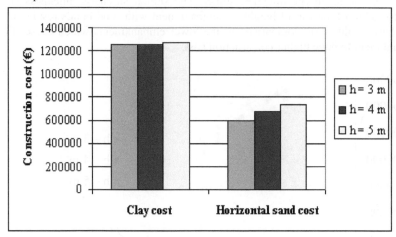

Figure 5.19: Effect of embankment height on the construction cost of embankment without sand bunds parts

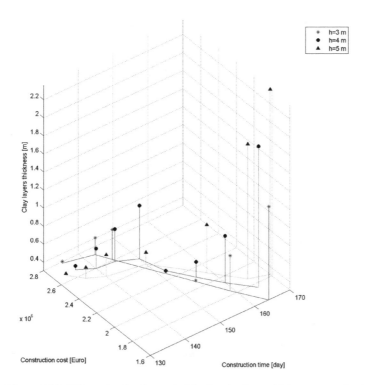

Figure 5.20: Effect of embankment height for clay layer thickness at optimum

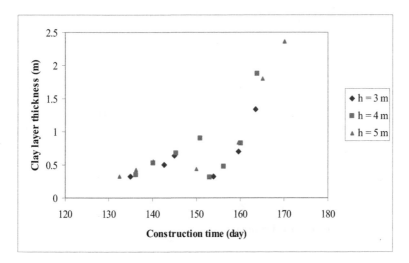

Figure 5.21: Relationship between construction time and clay layers thickness for various embankment heights

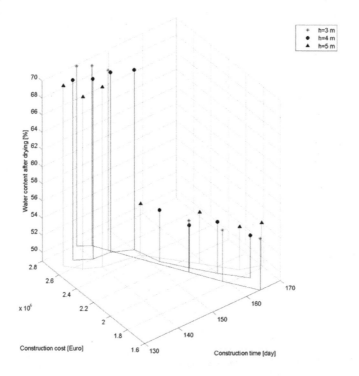

Figure 5.22: Effect of embankment height on water content after drying

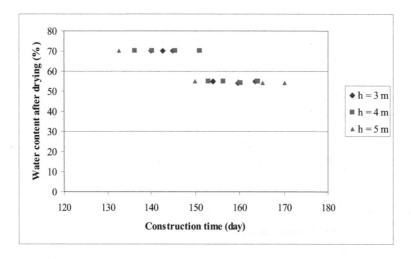

Figure 5.23: Relationship between construction time and water content after drying for various embankment heights

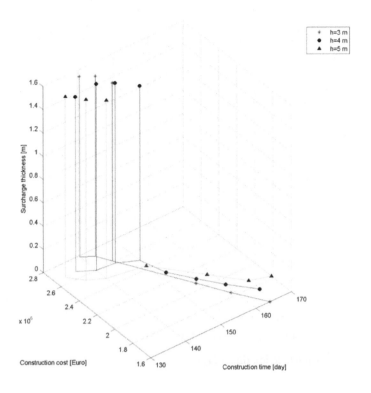

Figure 5.24: Effect of embankment height on surcharge thickness

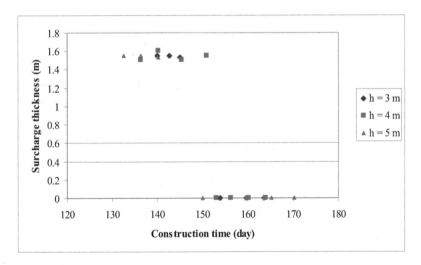

Figure 5.25: Relationship between construction time and surcharge thickness for various embankment heights

5.7.2 Sensitivity to changes in plasticity index

It appears from Figure 5.26 that changes for plasticity index affect the construction time. With low plasticity clay the design requirements can be achieved within a shorter construction time than high plasticity clay at nearly the same construction cost. Figures 5.27-5.32 show two different parts. The discontinuity of the graph is due to the change in method of construction of the embankment; with sand bunds for a short construction time and embankments without sand bunds for a long construction time. This change has influence on the Pareto frontier appearing as a sharp bend in the curve. At the same specific construction time low plasticity clay can be used in the construction with lower costs than high plasticity clay due to the lower number of clay layers required as shown in Figures 5.27 and 5.28.

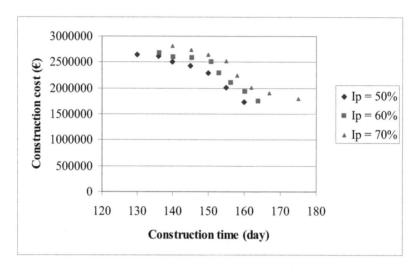

Figure 5.26: Effect of plasticity index on construction cost

The plasticity index has an effect on the required water content after drying, and therefore on workability, and clay layer thickness. The optimum results of water content after drying are shown in Figure 5.29 and 5.30. For the embankment with sand bunds, dryer clay has been used for $I_P = 50\%$ clay compare to $I_P = 60\%$ and 70% clay at the same construction time and gives lower construction cost. For the embankment without sand bunds, $I_C \approx 0.75$ is used for all plasticity indexes but lower plasticity index gives lower construction cost due to the thicker clay layers used. Figure 5.31 and 5.32 show the effect of plasticity on clay layer thickness. The construction time of the low plasticity clay is shorter than for high plasticity clay for the same clay layer thickness. Not included is the effect of the low I_P on the increased sensitivity to weather conditions. Low I_P clays tend to become easily unworkable already at moderate precipitation. These clays often require the use of additives to enable stable working conditions which is not included in the scope of this thesis.

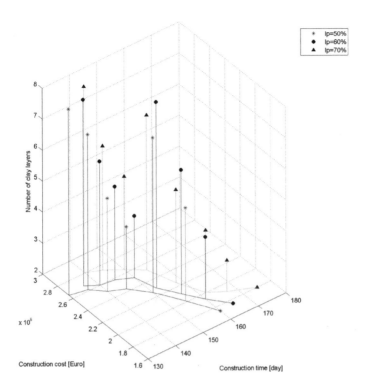

Figure 5.27: Effect of plasticity index for the number of clay layers at optimum

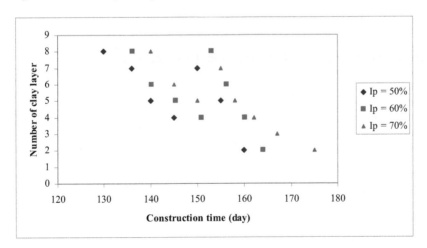

Figure 5.28: Relationship between construction time and number of clay layers for various plasticity indexes

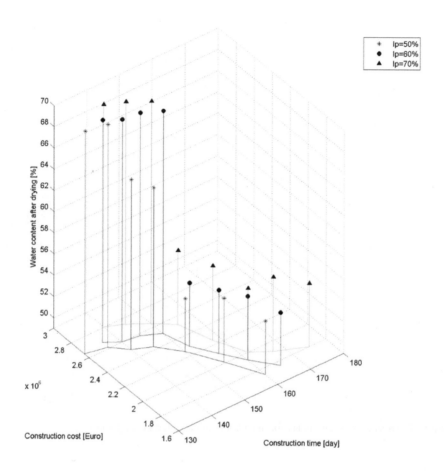

Figure 5.29: Effect of plasticity index for water content after drying at optimum

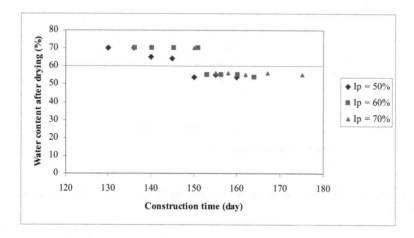

Figure 5.30: Relationship between construction time and water content after drying for various plasticity indexes

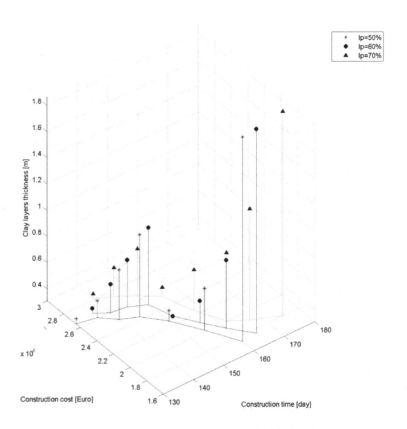

Figure 5.31: Effect of plasticity index for clay layer thickness at optimum

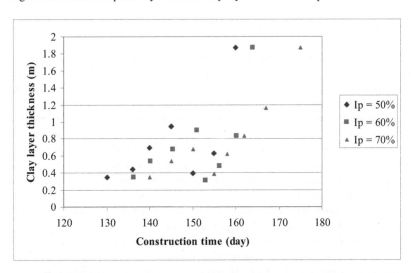

Figure 5.32: Relationship between construction time and clay layers thickness for various plasticity indexes

5.7.3   Sensitivity to unit costs

Figure 5.33 shows effects on the construction cost of embankments of the change of cost of horizontal sand layers and cost of clay. The change of cost of horizontal sand layers has a small effect for the selection of the type of embankments. The embankment with sand bunds is preferred when the construction time is less than 150 days and the embankment without sand bunds when the construction time is more than 150 days. This is the same as for the embankment constructed with normal cost for sand except that the costs are higher. With increased cost of clay, the selection of embankment type changes. The embankment with sand bunds is selected for all construction times. The construction cost is almost the same for all construction peroids.

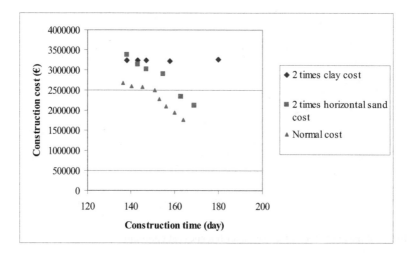

Figure 5.33: Effect of unit cost on construction cost

Figures 5.34 and 5.35 show the effect of changing of the unit costs for the optimum water content after drying. The change of costs of clay shows that wet clay is preferred over dry clay for all conditions. Consequently only embankments with sand bunds are selected for the construction instead of selecting embankment with sand bunds for a short construction time and embankment without sand bunds for a long construction time. The water content after drying of the clay decreases slightly when the construction time increases. The consistency index is 0.5 for the shortest possible construction time and increases to around 0.6 for the longest construction time.

When the construction time is less than 150 days the embankment with sand bunds is used for the construction. The change of cost of horizontal sand layers and the change of cost of clay does not have any effect on the optimum solutions, i.e. water content of clay after drying, number layer of clay, clay layer thickness, surcharge thickness and degree of consolidation compare to the optimum solutions of the normal cost embankment at the same specific construction time.

If the construction time is more than 150 days, the change in horizontal sand layer costs has no effect on the optimum solutions but the change in clay cost has effect on the type of the embankment and the optimum solutions has changed as shown in Figures 5.36-5.39.

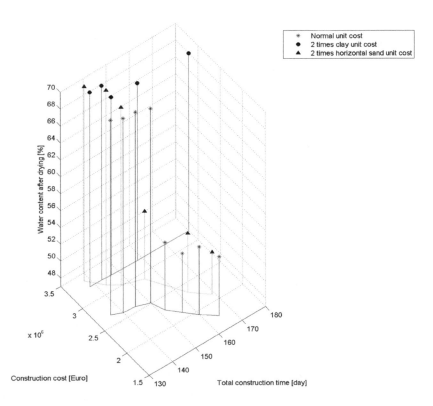

Figure 5.34: Effect of changing clay cost and horizontal sand cost for water content after drying at optimum

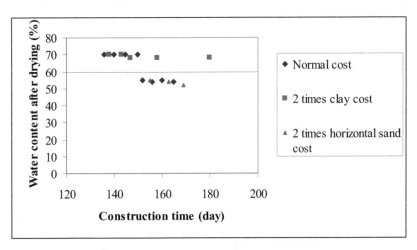

Figure 5.35: Relationship between construction time and water content after drying for changing of unit cost

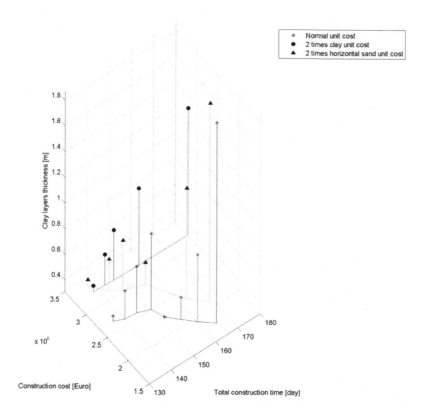

Figure 5.36: Effect of changing clay cost and horizontal sand cost for clay layer thickness at optimum

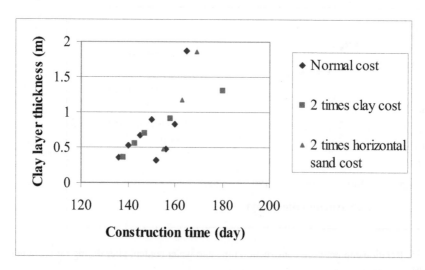

Figure 5.37: Relationship between construction time and clay layer thickness for changing of unit cost

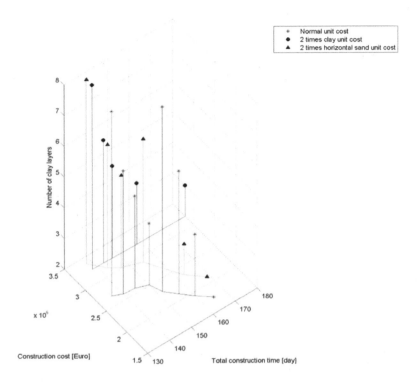

Figure 5.38: Effect of changing clay cost and horizontal sand cost for number of clay layers at optimum

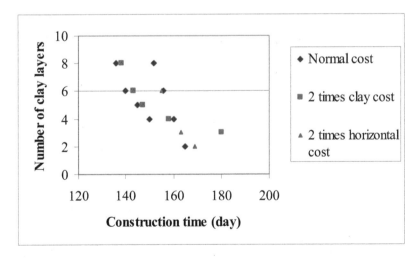

Figure 5.39: Relationship between construction time and number of clay layers for changing of unit cost

6
Field test

This chapter describes a test project of Rotterdam Port Authority that investigated the potential of reuse of unpolluted dredged material for embankment construction. The clay-sand layer or sandwich construction technique has been applied for a trial embankment in this project. The dredging sludge was ripened in ripening fields of the Slufter, Rotterdam, a large disposal site for dredging sludge and the clay was used for construction of the embankment. The monitoring results of this trial embankment were used to verify the values predicted by the ANN models in section 5.3 and to evaluate the potential of this technique to improve very soft organic soil for embankment construction. Results of field and laboratory investigations of physical and engineering properties relevant to the dredged material from the Slufter are presented. The study of the field behaviour of the trial embankment allows comparisons to be made between the computational models and full-scale behaviour. The results from the optimisation analysis and the investigation of the trial embankment are used to develop a method for design and construction of clay-sand layered embankment.

6.1   Test site

The site of the test embankment is located at the Slufter. Unpolluted dredged material has been kept separately from polluted material and was conditioned before it was used for construction of the trial embankment. The dewatering by evaporative drying of the dredged material was carried out at ripening fields at the Slufter. Figure 6.1 shows the location of the ripening fields consisting of five depots. Depot 1 has the lowest ground level and the highest ground water level, which caused very poor drainage conditions. Depot 2 and 3 are laid on coarse sand creating good drainage for these depots. Depot 4 and 5 are on clay again hampering drainage of these depots.

6.2   Characterisation and description of very soft organic clay at the Slufter

The materials studied in this experimental program are primarily fine-grained clayey silts. Samples were obtained from ripening fields of the Slufter. They are dark grey to black in colour and contain some organic matter. The available detailed test data are summarised in Table 6.1. A relationship of the undrained shear strength with water content was obtained from laboratory Fall Cone Tests on clay with $w_l = 95\%$, $w_P = 35\%$, and $I_P = 60\%$ is shown in Figure 6.2.

These laboratory data were used as soil parameters in the simulation programs that were used to develop ANN models as described in section 5.3 and for the RIJPING program to simulate drying behaviour of this very soft clay which is described in section 6.5.

6.3   Ripening

Fugro carried out the investigation of dewatering by evaporative drying in the laboratory and at the ripening fields under the supervision of Road and Hydraulic Engineering Institute, Directorate-General of Public Works and Water Management. The period of investigation was between October 2001 and September 2002. The investigation started in summer when the evaporation rate is higher than the precipitation. The dredged material was dumped at the ripening area since October 2001. The material was formed into bunds as shown in Figure 6.3 when it was strong enough. The average height of bunds was 1.5 m. A trench between the bunds assisted in dewatering the soil masses in the bunds by providing rapid drainage of precipitation. The precipitation flows through the desiccation cracks, into the trenches, and is removed. From time to time a hydraulic excavator was used to agitate the material for acceleration of the drying process.

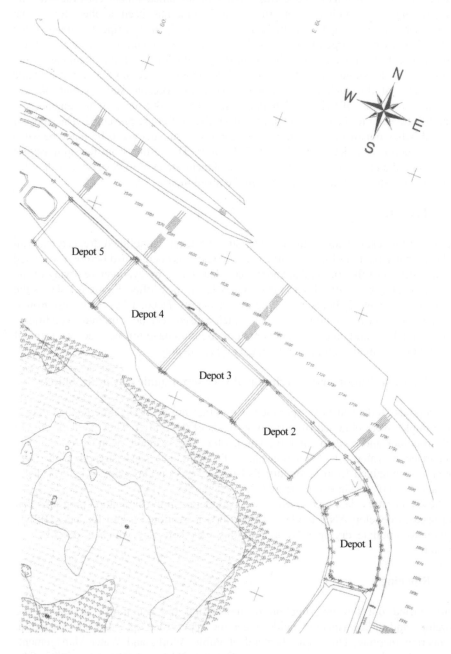

Figure 6.1: Layout of ripening fields at the Slufter

Table 6.1. Soil properties

Parameter	Range value	Unit
Particle size		
> 63 μm	8.7 - 18.5	%
< 16 μm	50.2 - 74.4	%
< 2 μm	22.6 - 33.4	%
Atterberg limit		
- Liquid limit (w_L)	79 - 117	%
- Plastic limit (w_P)	30 - 42	%
- Shrinkage limit (w_S)	27 -33	%
- Plasticity index (I_P)	47 - 75	%
Bulk density (γ)	1.3 - 1.4	g/cm^3
Solid density (γ_p)	2.54 - 2.62	g/cm^3
Organic content	7 - 11	%
Strength parameters		
- Effective cohesion (c')	6 - 8	kPa
- Effective angle of shearing resistance (ϕ')	30	degree
Consolidation parameters		
- General compression constant below p_g (C)	26 - 59.5	
- General compression constant above p_g (C')		
- Pre-consolidation pressure (p_g)	8 - 13.8	
- Primary compression index below p_g (C_p)		
- Secondary compression index below p_g (C_s)	17 - 28	kN/m^2
- Primary compression index above p_g (C'_p)	18 - 168.6	
- Secondary compression index above p_g (C'_s)	87 - 1.746E+03	
- Vertical coefficient of consolidation (C_v)		
- Secondary compression index (C_α)	11 - 18.8	
- Vertical coefficient of volume compressibility (m_v)	79 - 192.1	
- Permeability coefficient (k)	9.7E-10 - 4.6E-07	m^2/s
	3.9E-03 - 7.2E-03	
	1.2E-01 - 7.25E-01	m^2/MN
	4.3E-11 - 5.9E-9	m/s
Soil-water characteristics		
- Residual water content (θ_r)	0.100 - 0.358	
- Saturated water content (θ_s)	0.613 - 0.677	
- Van Genuchten coefficient (α)	0.005 - 0.022	cm^{-1}
- Van Genuchten coefficient (n)	1.095 - 1.200	
- Saturated hydraulic conductivity (K_s)	2 - 14	cm/day

The result of the drying from 19-06-2002 to 23-08-2002 in the Fugro laboratory is shown in Figures 6.4 and 6.5 for a mixed sample of depot 1 and 2 and a mixed sample of depot 3-5 respectively. The average water content of the clay on 19-06-2002 was 83% for depot 1 and 2 and 104% for depot 3 to 5. The average water content of the clay on 23-08-2002 was 36% for depot 1 and 2 and 43% for depot 3 to 5. The average water content at the ripening fields was determined 3 times from 21-05-2002 to 5-09-2002 as shown in Figure 6.6. It can be seen that the average actual water content in the field is higher than in the laboratory. However, if the favourable condition for drying has been met such as in depot 2 and 3 the drying to a low water content in the field can be achieved. The discrepancies between field observation and drying simulation with the program RIJPING were too large to be taken as approximation of actual conditions. The relationship between average water content and drying time in the field was used in the optimisation model (Appendix J) to make the model more practical.

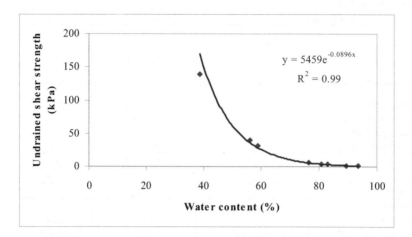

Figure 6.2: Water content – undrained shear strength relationship

Figure 6.3: Clay in ripening field

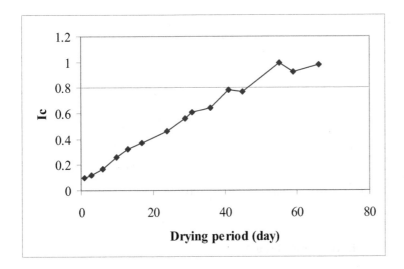

Figure 6.4: Consistency index of drying sample in laboratory (depot 1 and 2) during 19-06-2002 – 23-08-2002

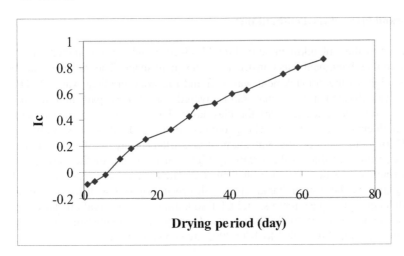

Figure 6.5: Consistency index of drying sample in laboratory (depot 3, 4 and 5) during 19-06-2002 – 23-08-2002

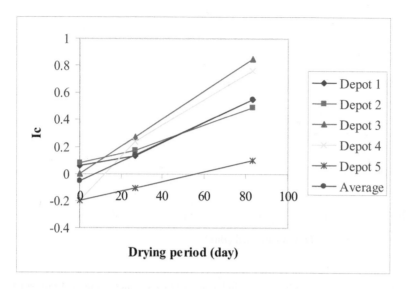

Figure 6.6: Consistency index of drying sample in ripening field during 21-05-2002 – 13-08-2002

6.4   Clay-sand layered trial embankment

The construction of the embankment started on 27-03-2003 and was completed on 12-05-2003. The trial embankment was constructed in a square shape. The top size was 50 × 50 m. The height of the embankment was 4 m and the slope gradient was 1:2. The embankment was divided into a clay-sand layered (sandwich) part and a clay monolayer part. For the sandwich part the clay layer thickness was 1.5 m and sand drainage layer thickness was 0.5 m. The geometry of the embankment is shown in Figure 6.7. During construction the embankment was lifted in layers of 30 cm thick. Compaction was difficult due to the relatively high water content of the clay. Field conditions during construction are given in Table 6.2. The layout and a cross section of the instrumentation of the trial embankment are shown in Figure 6.8. Instrumentation included settlement plates, piezometers, and CPT sounding. The observation started on 17-04-2003 and ended on 4-09-2003. The results of settlement, pore pressure and water content monitoring from 17-04-2003 to 4-09-2003 are shown in Table 6.3.

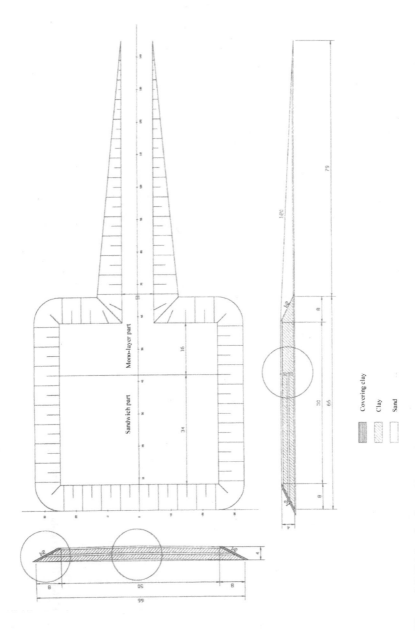

Figure 6.7: Trial embankment geometry

Very soft organic clay applied for road embankment

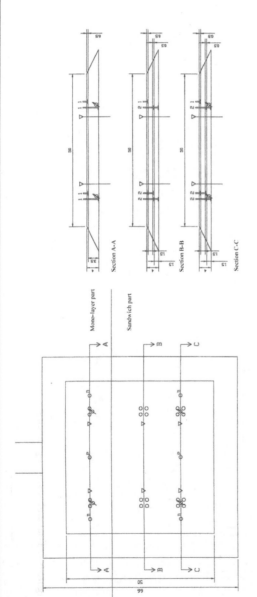

Section A-A

Section B-B

Section C-C

Mono-layer part

Sandwich part

Symbol

⊥ ○   Settlement plate

∅   Piezometer

▽   Sounding

○ᴰ   Density test

○ⱽ&ᴰ   Volume and density test

○ᴾ   Proctor test

Figure 6.8: Instrumentation layout

Table 6.2. Field conditions during construction

Material	Thickness (m)	Water content (%)	Dry density (kN/m³)	Bulk density (kN/m³)
Top sand layer	0.50	6.6	16.1	17.0
Second clay layer (Upper layer)	1.50	57.1	10.1	15.8
Second sand layer	0.50	6.2	15.9	16.9
First clay layer (Lower layer)	1.50	55.9	10.2	15.9
First sand layer (base)	0.50	8.3	15.8	17.1

Table 6.3. Monitoring results

Property		Mono-layer		Sandwich	
		Upper layer	Lower layer	Upper layer	Lower layer
Bulk density (kN/m³)		15.7	15.7	15.6	16.0
Dry density (kN/m³)		10.3	10.5	9.9	10.7
Water content (%)		53.4	49.6	58.6	49.5
Porosity (%)		60.7	59.6	62.1	58.9
Strength parameter - Effective cohesion (c') (°)		33	32	28	32
- Effective angle of shearing resistance (ϕ') (kPa)		0	0	0	3
Consolidation parameter - General compression constant below p_g (C)		13.7	19	32.1	23.7
- General compression constant above p_g (C')		8.1	8.5	10.5	8.6
- Pre-consolidation pressure (p_g) (kN/m²)		25	27	29	28
- Vertical coefficient of consolidation (C_v) (m²/s)		1.1E-07	2.5E-08	6.4E-08	6.3E-08
Reduction of pore water pressure after completion of the construction (h)		0.10	0.55	0.85	1.00
Settlement after completion of the construction (m)		0.03		0.05	

6.5   Analysis of the monitoring results

*   *Evaporative drying*

The simulation program for ripening soil RIJPING was run with soil parameters obtained from laboratory testing (Table 6.1). Input data on the weather conditions were as shown in Figure 6.9, consisted of average actual precipitation and potential evaporation from 1992 till 2001 at De Bilt. This data set is considered to represent the average values of precipitation and evaporation in the Netherlands. The program RIJPING was run with the soil water retention curve, hydraulic conductivity curve of the clay and the necessary information on the drainage situation and weather data. The detail of RIJPING is given in Appendix I.

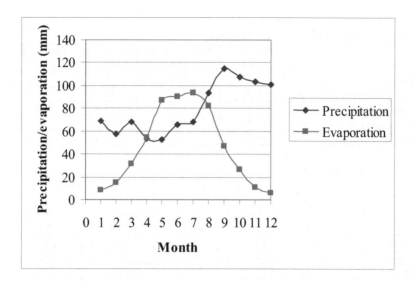

Figure 6.9: Average monthly precipitation and evaporation

The calculation results of the RIJPING were compared with the monitoring result and are shown in Figure 6.10. The water content obtained from the monitoring is less than the water content obtained from RIJPING (average water content for 0.50 m thickness). This may be due to the agitation and stocking of the clay in bunds, which accelerates the evaporation process, and RIJPING does not include this agitation in its model. Some researchers have mentioned accelerating the drying process by mechanical conditioning. They found that at the same drying period the water content of conditioned soil was less than non-conditioned soil (Krizek et al., 1973; Krizek et al., 1978; Thomas, 1990; Benson Jr. and Sill, 1991; Thomas, 1993). The mechanical conditioning can be very effective for dewatering wet material if the process of agitating is properly scheduled. The effectiveness of mechanical conditioning depends on many factors such as mixing schedule, evaporation rate, drainage condition, and layer thickness of the material. RIJPING was run with data from Thomas (1990) to compare the result. The test result of the Stepps Field Dewatering Trials is used because of its similarity of soil and field condition with the present study. The material used in these trials was material dredged from the River Clyde, and is clayey silt. It was placed in an observation tank with initial water content of 106% and thickness of 0.50 m. Agitation was not applied during the test. Monitoring of the water content was undertaken at regular intervals. The monitoring period was 100 days. The comparison is shown in Figure 6.11. It can be seen that the result obtained from RIJPING agrees well with the monitoring result, but the water content obtained from RIJPING is still somewhat higher than the observed values. However, the difference is small and acceptable. Therefore, RIJPING has the ability to predict the water content of clay during dewatering by evaporative drying but agitating makes the water content significantly lower than the simulation program prediction.

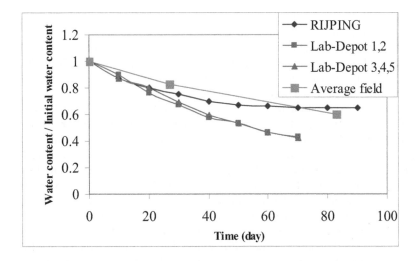

Figure 6.10: Comparison of water content from RIJPING and monitoring results at Slufter

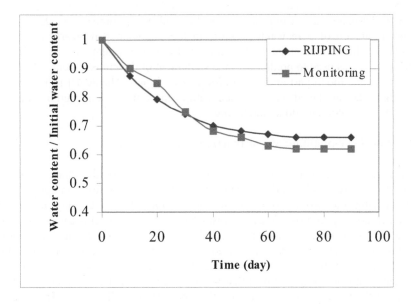

Figure 6.11: Comparison of water content from RIJPING and monitoring results at Stepps Field

*   *Clay-sand layered embankment*

Tables 6.1 and 6.2 show that, the bulk density of clay during construction increases due to the compaction by bulldozer operation. The maximum dry density of 13.3 kN/m^3 cannot be achieved with the bulldozer. About 75% of the maximum dry density can be induced by bulldozer compaction for the $I_C = 0.7$ clay. Embankment stability was no problem during construction. The slope stability ANN model was used to predict the embankment stability to verify the model. The computation results show that the slope stability of the embankment (F.S = 2) can be maintained for a maximum top sand thickness of 1.88 m for the sandwich part and 1.77 m for the mono clay layer part. This agrees with the field observation that the embankment was stable for the top sand thickness of 0.50 m. Tables 6.2 and 6.3 show that there were no significant changes in bulk and dry density of the clay layers after 3 months after completing the construction. But there is a noticeable reduction of the water content and porosity in the lower clay layer for both mono part and sandwich part due to expulsion of water by the overburden. Only, a small change takes place in the upper clay layer due to the lower overburden. This is in line with the expectation of an $I_c \approx 0.7$ in The Netherlands for a longer term equilibrium water content. Strength and consolidation parameters are still in the range that was determined before the construction started. Thus, within 3 months there is no significant improvement of those parameters. For the pore water pressure there is a significant reduction in the sandwich part compared to the mono clay layer part. The reduction was higher in the lower layer due to the higher overburden pressure. The ANN model for settlement time was used to predict the embankment settlement time monitored in the field. For the mono clay layer part the predicted results were 73 days for 0.03 m settlement and 38 days for 0.05 m of the sandwich part. There are some differences between monitoring and predicted results. The model prediction was more similar to the observed values for the mono clay layer part. The less accurate result for the sandwich part may be due to complexity of the clay-sand layer behaviour and lesser efficiency of the clay-sand layer to dissipate water from the clay in the field when compared to theory. The higher reduction of pore water pressure in the sandwich part clearly shows the benefits of the horizontal sand layer in accelerating consolidation. The settlement ANN model was used to predict the total settlement of the embankment, which was 0.15 m for the mono part and 0.17 m for sandwich part. The actual results can be less than the predicted values since settlement during construction was not included in the computational modelling.

6.6   Optimising embankment construction using wet clay

Clay is recognized as problematic material for construction of road embankments. The presence of wet clay fills in road embankment construction may result in considerable problems during the works, often with cost implications. Wet clay cannot be compacted to the required dry density, causes trafficability problems for construction equipment, takes longer construction time, and may cause excessive settlement during service time. However, wet clay can be used in road embankment construction using a clay-sand layered construction. Combination of this technique with surcharges for road embankment construction has proved that wet clay can be used as fill material. But a consistent method for optimising use of this material is still absent. This section presents a rational method for conditioning wet clay, a rational design method, and a rational method for construction. The optimum thickness of clay and sand layers is presented here. The design method is based on minimising construction cost while providing a suitable formation thickness satisfies all requirements. The design method is intended to achieve a stable road embankment that fulfils all design requirements such as settlement amount, stability, final height, and construction time with the minimum construction cost.

6.6.1 Conditioning wet clay

This section describes an example how the simulation program RIJPING may be used in practice. The example considers the case of the soil at the Slufter proposed for use as an earthwork fill. As described earlier initially this soil is unsuitable due to its very

high water content. The average water content is 83% relating to an I_C of 0.28. If clay with $I_C = 0.70$ (water content of 58%) is needed for embankment construction, answers to the following questions are required:
* Can the soil be dewatered within a reasonable time period and how long will drying take?
* What is the proper layer thickness for drying?
* What are seasonal effects?

To answer those questions dewatering calculations have to be performed. Calculations were performed using RIJPING for the months in which the evaporation rate is higher than the precipitation rate (April–July). The required drying periods for dewatering clay with $I_C = 0.28$ to $I_C = 0.70$ for clay layer thickness of 0.10, 0.25, and 0.50 m are tabulated in Table 6.4 and are plotted in Figure 6.12.

Table 6.4. Required drying period

Starting month	Drying period (day)		
	0.10 m	0.25 m	0.50 m
April	43	54	90
May	20	28	65
June	25	30	70
July	25	31	70

Decision can be made using the graph, which is given in Figure 6.12. By inspection, the optimum initial layer thickness is about 0.25 m. The layer thickness of 0.10 m is obviously not economical. The most effective drying period is between May and July. A layer thickness of more than 0.50 m is not efficient. The graph presented does not include the effect of agitation, a quicker drying time and dewatering may be achieved with the assistance of agitation. However, for a layer thickness of 0.25 m this will have little extra effect since the surface of the bunds dries quickest. Good management practices and quality control are essential during dewatering by evaporative drying. These include prevention of waterlogging by providing good drainage and providing sloping and open drainage trenches across the drying area can assist surface drainage.

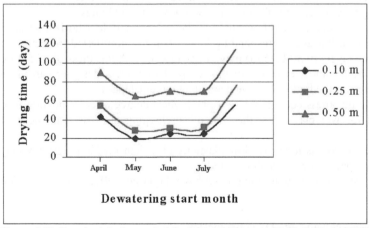

Figure 6.12: Required drying period for initial thickness of 0.10, 0.25, and 0.50 m.

6.6.2   Clay-sand layer embankment design and construction

Optimisation of the design of a clay-sand layered embankment is necessary because the construction cost depends on the design criteria selected and incorrect selections can significantly increase construction costs. The following design considerations indicate the requirements, which must be met if safety and proper economy are to be assured:

*   The primary design concern for a road embankment is stability and settlement. Stability has to be maintained during construction and after construction. Large settlement during road service can damage the pavement structure resulting in a reduction of the serviceability. Prediction of settlements is necessary to quantify in advance the amount of fill material required, and the duration of construction, and post-construction settlement. Availability of accurate methods to evaluate the effects of stability and settlement is a prerequisite to enable a design optimisation with respect to construction and maintenance costs;
*   Using very soft clay as a construction material can cause a large settlement and a long settlement time. It can be necessary to apply horizontal sand layers and a preloading to the embankment to meet settlement time requirement. Horizontal sand layers can be applied to accelerate the consolidation process and to help to improve the discharge of excess pore pressures in the embankment. An additional surcharge fill can be placed temporarily so that sufficient settlement can be reached sooner than with the normal fill procedure. After removal of the surcharge load further settlement will be minimised;
*   A trade-off curve of construction cost and time should be constructed to assist a designer to make a proper decision in design;
*   For each specific construction time the proper clay water content, number of clay layers, clay thickness, and surcharge thickness has to be selected in order to yield minimum construction cost while satisfying all constraints;
*   Wet clay (I_C = 0.5-0.7) can be used for a short construction time, but it can cause high construction cost. Clay dried from wet condition to water content with $I_C >$ 0.7 requires a longer construction period but lower costs. In general the direct construction cost reduces if the construction time increases;
*   For wet clay, an I_C = 0.50-0.58 is an optimum value for construction. Embankments with sand bunds are recommended for such wet clays. I_C = 0.75 is the optimum value for clay for the embankment without sand bunds;
*   A reduction of direct construction cost can be achieved if less horizontal sand layers are applied but at a corresponding longer construction period. Selection of many thin clay layers (0.30-0.75 m) may not be effective. A thickness of 0.75-1.00 m seems to be the most effective in this study. However, a designer can make decisions based on the trade-off curve depending circumstances;
*   The surcharge thickness corresponding to the highest degree of consolidation is to be preferred for wet clay when the cost of surcharging are less than those of conditioning;
*   The construction cost is lower for low plasticity clay than for high plasticity clay within a same specified construction period;
*   Changes in cost of clay (conditioning cost) have the largest effect on construction cost.

6.6.3   General remarks on wet clay construction

The use of wet clay requires placing and compacting techniques different from those normally employed. As the construction proceeds and if softer material is used, there will be an increasing problem with placement and compaction of the clay. The normal rubber tyre plant is unable to operate over wet clay. The use of machinery on tracks is preferred for this clay. Excavation has to be carried out by track-type excavators or front loaders. The wet clay is unable to support construction traffic and suitable haul roads therefore have to be provided for. Bulldozers provided with extra large tracks if necessary can effect spreading. The material may be dumped and bulldozers may push the clay on to the fill and compact it in layers.

Clay with $I_C \approx 0.70$ can form a stable embankment with adequate workability. The compaction difficulty due to the low bearing capacity of the clay and variation of clay water content causes uncontrollable clay layer density. Lifts < 0.3 m are to be used and compaction with tired rollers or flat rollers is generally recommended. Compaction by bulldozer operations can induce about 75% of the maximum dry density for lifts of about 0.25 m thick. Over compaction can induce excess pore pressures in the clay with a consequent decrease in stability. A criterion for compaction based on the maximum dry density and the optimum water content cannot be applied to very soft clay since the water content during operation is much higher than the water content that gives the required dry density. To control the construction the number of bulldozer passes on the clay should be specified. According to past experience only one or two passes of bulldozers appear adequate.

To avoid differential settlements within embankments, the clay is placed in layers of even thickness and ruts are progressively filled before the placement of the next layer. Adequate slope for runoff drainage during construction must be provided to reduce the amount of water that can penetrate into the clay layer.

7
Main findings and outlook

Very soft organic clay has characteristically a very high water content, very low strength and very high compressibility. It becomes a more serious problem due to the increasing amount of this clay becoming available from development projects for which there is little disposal area available. The re-use of this clay as embankment fill material may be a way to solve at least part of the problem. The goal of this study was optimisation of the use of very soft organic clay as a fill material for road embankment construction. The optimisation is based on a rational method for conditioning and emplacing very soft organic clay in a safe and economic way. For using very soft organic clay for road embankment at economic construction cost, an optimisation technique was applied. An equipment selection model and a multi-objective optimisation model for clay-sand layered embankment construction were developed for this particular problem. A Genetic Algorithm (GA) was applied as an optimisation technique to find optimum solutions to the problem. The main results from the study are :

* It is necessary to improve the engineering properties of very soft organic clay before using the clay as a fill material. Water content is considered to be the most important factor in determining the usefulness of the clay. Reduction of the water content, as much as feasible and practical, is the most significant action in conditioning the clay to make it suitable for embankment construction;

* Dewatering by evaporative drying, dewatering using horizontal sand drainage (clay-sand layered or sandwich technique) and pre-loading (surcharging) are considered to be simple, practical and cost-effective means for this improvement;

* Results from simulation computer programs (FSCONBAG for consolidation simulation, MStab for slope stability calculation, RIJPING for evaporative drying simulation) were used to simulate the behaviour of very soft organic clay. But, instead of directly linking these programs to the optimisation model. Artificial Neural Network (ANN) models were developed to predict settlement amount, settlement time, and stability for clay-sand layered embankments in order to reduce the computational load during optimization. The relations of parameters that influence clay behaviour were determined from the simulation results by ANN due to its renowned ability to solve complex pattern recognition problems. These ANN models were integrated in the clay-sand layered optimisation model instead of simulation models with the aim to reduce complexity of input data and computation time during the optimisation calculation process. The developed ANN models have shown their high accuracy of predicting results and could be succesfully integrated in the optimization procedure. The computer models did, however, not give very accurate predictions of the modelled phenomena. In view of the scope of the subject of this thesis the results were adequate and also will be for genral practical applications, since the predictions were rather conservative.;

* For using very soft organic clay for road embankment at economic construction cost, an optimisation technique was essential. Optimisation models were developed for this particular problem. These optimisation models consist of an equipment selection model and a multi-objective optimisation model for clay-sand layered embankment construction. A Genetic Algorithm (GA) was applied as optimisation technique to find the optimum solution of the problem. GA was selected due to many benefits over other techniques such as its robustness in solving many type of

engineering problems, its solutions are likely to be global optimums, its flexibility to modify for further development models. The equipment selection model is a model for selection of size and number of the earthmoving equipment fleet, yielding the minimum equipment costs. The results from this model were used as parameters to develop the cost models that are a part of the multi-objective clay-sand layered construction model. The optimum results show that the minimum equipment cost can be found when the production rate is more than 200 m³/hr. A smaller number of large sized equipment is preferred to be selected rather than a large number of small size equipment for the same production rate. Among the construction materials wet clay ($I_C < 0.70$) has the highest equipment cost;

*   A computer program was developed to integrate the clay-sand layered embankment design in the GA optimisation process that was applied to find the solutions for the multi-objective optimisation model. The relationship of drying behaviour parameters was obtained from field monitoring results at Slufter drying field. The relationships of settlement behaviours and stability were obtained using ANN models. The formulation of the optimisation model considering direct construction costs and construction time as objective functions was developed. Minimising construction cost and construction time when all constraints are satisfied, was the objective of the optimisation. By using the so-called ε-constraint or trade-off approach, the computer program could perform the optimisation process. The algorithm was tested on several problems. The results indicated that the optimisation model and the GA can be used efficiently in finding the set of the optimum solutions to construct a so-called Pareto frontier or trade-off curve. A proper decision for clay-sand layered embankment construction can be obtained by the assistance of this trade-off curve;

*   The results of the optimisation analysis are as follows:
    1. The relationship between the two objectives is not linear. The direct construction cost decreases when construction time increases. The construction cost increases rapidly when a short construction period is required;
    2. For a short construction period, the embankment with sand bunds with high water content clays and many horizontal sand layers is preferable. When the construction time is longer the dryer clay is chosen to construct embankments without sand bunds. For both types of embankment a reduction of horizontal sand layers causes a reduction in construction cost;
    3. For thin clay layers (0.30-0.75 m) the construction cost increases with a higher rate than for thick layers (0.75-2.40 m);
    4. For a specific construction time the number of clay layers, clay layers thickness, and surcharge thickness has to be chosen properly in order to achieve all constraints and yield minimum construction cost;
    5. Construction cost, construction time, and design variables can change if embankment dimension, clay properties, and unit costs of embankment construction are changed. The Trade-off curve for specific conditions can be constructed with the same approach used in this study to assist a decision maker to make a proper decision for that condition;

*   A trial clay-sand layered embankment was constructed at the Slufter, Rotterdam in order to investigate the potential of using very soft organic clay for embankment construction. The monitoring results have been used to verify the models that used to predict the clay behaviours in the optimisation model. The findings of the monitoring are:
    1. The investigation of dewatering by evaporative drying was carried out in laboratory and at ripening fields. The dewatering rate in the laboratory was

higher than in the fields. The final $I_C = 1$ and 0.85 for good and poor drainage conditions respectively can be achieved in the laboratory for 65 days drying period. In the field average $I_C = 0.65$ can be achieved within 82 days of drying. These results were compared with the results that were calculated using the simulation program for drying soil RIJPING. It was found that the program gave too high water contents ($I_C = 0.62$) for the drying periods. To make the optimisation results most practical the relationship of drying parameters, which was used in the optimisation model, was taken from the monitoring results from the fields;

2. The investigation of embankment behaviour has shown that about 75% of the maximum dry density can be gained from bulldozer operation. Clay with an average I_C of 0.70 can be used to construct an embankment of 4 m height with adequate workability and without problem of stability. A noticeable reduction of water content, porosity, and pore water pressure in the embankment has occurred. The pore water pressure there was a significantly reduced in the sandwich part compared to the mono clay layer part. The sandwich part also had a higher settlement than the mono clay layer part. The monitoring results were compared with the results obtained from ANN models used to predict embankment stability, settlement amount and settlement time of the embankment to verify those models. The finding was that the slope stability ANN model was in agreement with the monitoring result in that the slopes did remain stable during the monitoring period; the ANN models for settlement gave a faster settlement time than the monitoring. The models gave more accurate results for the mono clay layer part. The less accurate results for the sandwich part may result from the complexity of clay-sand layered behaviour and the reduced efficiency of sand layers to dissipate water from the clay in the field compared with theory;

* The results from the optimisation analysis and the investigation of the trial embankment were used to come to design and construction recommendations for use of very soft organic clay for road embankment construction. Recommendations for using very soft organic clay for road embankment construction are as follows:

- *Conditioning wet clay*
  1. Good drainage conditions such as low ground water level, providing sand base and adequate drainage trenches in ripening fields can assist clay dewatering in the ripening fields;
  2. The thickness of drying clay and the agitation play an important role for the efficiency of evaporative drying. Agitation of the clay in ripening fields can be very effective for dewatering if the process of agitation is properly scheduled.

- *Clay-sand layered embankment design and construction*
  1. The primary design concern for clay-sand layered embankment is stability and settlement;
  2. The clay-sand layered technique can accelerate the consolidation process. Surcharging can reduce post construction settlement. These techniques should be applied to an embankment that uses very soft organic clay as a construction material;
  3. A trade-off curve of construction cost and time can be constructed to assist a designer to make a proper decision in design;
  4. Clay with $I_C = 0.5$-0.58 can be used to construct a clay-sand layered embankment but workability and stability are factors causing high construction cost. Dryer clay, $I_C = 0.75$ can be used to construct at lower

cost;

5. A reduction of construction cost can be achieved if less horizontal sand layers, corresponding with longer construction time, are applied in a clay-sand layered embankment;

6. For a specific construction time, a proper clay water content after drying, the number of clay layers, the clay layer thickness, the surcharge thickness, and the degree of consolidation has to be selected in order to yield minimum construction cost and satisfy all constraints;

7. The use of wet clay requires different placing and compacting techniques to those normally employed. The use of machinery on very wide tracks is preferred;

8. The $I_C = 0.7$ clay can be used to construct a 4 m embankment height with adequate workability and stability.

In this study, very soft organic clay that causes many problems for embankment construction in deltaic areas was re-used as embankment fill material. Dewatering by evaporative drying, dewatering by horizontal sand drainage (clay-sand layers in a sandwich construction) and pre-loading (surcharging) were considered to be the most effective methods that can be applied for this type of soil. Artificial Neural Networks (ANN) and Genetic Algorithms can efficiently be used for this particular problem. By these means the use of very soft organic clay as a fill material for road embankment construction can be achieved in a safe and economic way. More data on such kind of construction would be an advantage in applying these models and methods in practice.

References

Abebe, A.J., and Solomatine, D.P. (1998), *Application of global optimization to the design of pipe networks*, Proc. 3 rd International Conference on Hydroinformatics, Copenhagen, pp. 989-995.

Abu-Hejleh, A.N., and Znidarcic, D. (1995), *Desiccation Theory for Soft Cohesive Soils*, Journal of Geotechnical Engineering, ASCE, Vol. 121, No. 6, pp. 493-502.

Alonso, E.E., Gens, A., and Lloret, A. (2000), *Precompression Design for Secondary Settlement Reduction*, Geotechnique, Vol. 50, No. 6, pp. 645-656.

Amirkhanian, S.N., and Baker, N.J. (1992), *Expert system for equipment selection for earth-moving operations*, Journal of Construction Engineering and Management, ASCE, Vol. 118, No. 2, pp. 318-331.

Arrowsmith, E.J. (1978), *Roadwork Fills-A Material Engineer's Viewpoint*, Clay Fills, Institution of Civil Engineers, London, pp. 25-36.

Back, T., Fogel, D.B., Whitley, D., and Angeline, P.J. (2000), *Mutation Operators*, Evolutionary Computation 1, Institute of Physics Publishing, Bristol and Philadelphia, pp. 237-255.

Belegundu, A.D., and Chandrupatla, T.R. (1999), *Optimization Concepts and Applications in Engineering*, Prentice-Hall, New Jersey.

Benson Jr., R.E., and Sill, B.L. (1991), *Evaporative Drying of Dredged Material*, Journal of Waterway, Port, Coastal and Ocean Engineering, ASCE, Vol. 117, No. 3, pp. 216-234.

Berger, D.J., and Tryon, S.O. (1999), *Approach to Designing Structural Slurry Walls*, Journal of Geotechnical and Geoenvironmental Engineering, ASCE, Vol. 125, No. 11, pp. 1011-1022.

Berggren, B.S.D. (1999), *Ground Improvement – an Overview*, Geotechnical Engineering for Transportation Infrastructure, Balkema, Rotterdam, pp. 1411-1418.

Black, W., and Lister, N.W. (1978), *The Strength of Clay Fill Subgrades: Its Prediction and Relation to Road Performance*, Clay Fills, Institution of Civil Engineers, London, pp. 37-48.

Boels, D., and Oostindië, K. (1991), *Rijping – A Simulation Model for Calculation of Water Balance, Ripening, Crack Formation and Surface Subsidence of Clay Sludge*, Manual, Version 1.0, The Winand Staring Center, Wageningen.

Boman, P., and Broms, B.B. (1978), *Behaviour of a Road Embankment Constructed of Soft Clay and Provided with Drain Strips*, Clay Fills, Institution of Civil Engineers, London, pp. 49-56.

Bowles, J.E. (1979), *Physical and Geotechnical Properties of Soils*, McGraw-Hill, Tokyo.

Bronswijk, J.J.B. (1988), *Modelling of Waterbalance, Cracking and Subsidence of Clay Soils*, Journal of Hydrology, 97, pp. 199-212.

Bronswijk, J.J.B. (1989), *Prediction of Actual Cracking and Subsidence in Clay Soils*, Soil Science, 148, pp. 87-93.

Budhu, M. (2000), *Soil Mechanics and Foundations*, John Wiley & Sons, New York.

Cargill, K.W. (1984), *Prediction of Consolidation of Very Soft Soil*, Journal of Geotechnical Engineering, ASCE, Vol. 110, No. 6, pp. 775-795.

Carrier, W.D., Bromwell, L.G., and Somogyi, F. (1983), *Design Capacity of Slurried Mineral Waste Ponds*, Journal of Geotechnical Engineering, ASCE, Vol. 109, No. 5, pp. 699-716.

Caterpillar. (1995), *Caterpillar Performance Handbook, Edition 26*, Caterpillar Inc., Peoria.

Charles, J.A., and Watts, K.S. (2001), *Building on Fill: Geotechnical Aspects*, Construction Research Communications, London.

Chong, E.K.P., and Żak, S.H. (1996), *An Introduction to Optimization*, John Wiley & Sons, New York.

Chow, Y.K., and Thevendran, V. (1987), *Optimisation of Pile Groups*, Computers and Geotechnics, Vol. 4, No. 1, pp. 43-58.

CUR (Centre for Civil Engineering Research and Codes). (1996), *Building on Soft Soils*, A.A Balkema, Rotterdam.

Day, R.W., and Axten, G.W. (1990), *Softening of Fill Slopes due to Moisture Infiltration*, Journal of Geotechnical Engineering, ASCE, Vol. 116, No. 9, pp. 1424-1427.

Dennehy, J.P. (1978), *The Remoulded Undrained Shear Strength of Cohesive Soils and Its Influence on the Suitability of Embankment Fill*, Clay Fills, Institution of Civil Engineers, London, pp. 87-94.

Deutekorn, J.R., Calle, E.O., and Termaat, R.J. (1997), *Economical Optimization of Soil Investigation,* International conference on soil mechanics and foundation engineering 14, Vol. 1, pp. 469-472.

Dohaney, W.J., and Forde, M.C. (1978), *Maximum Moisture Contents of Highway Embankments,* Clay Fills, Institution of Civil Engineers, London, pp. 95-100.

Dumbleton, M.J., and Burford, D. (1978), *Discussions: Construction, Placement and Methods of Treatment of Clay Fills,* Clay Fills, Institution of Civil Engineers, London, pp. 247-282.

Fujiyasu, Y., Fahey, M., and Newson, T. (2000), *Field Investigation of Evaporation from Freshwater Tailings,* Journal of Geotechnical and Geoenvironmental Engineering, ASCE, Vol. 126, No. 6, pp. 556-567.

Fukazawa, E., and Kurihara, H. (1991), *Estimation of Long-term Settlement for Soft clay improved by Preloading Method,* GEO-COAST'91, Vol.1, Yokohama, pp. 183-186.

Gen, M., and Cheng, R. (2000), *Genetic Algorithms and Engineering Design,* John Wiley & Sons, New York.

GeoDelft (2003), *MStab Version 9.7,* DelftGeoSystems, Delft.

Gibson, R.E., England, G.L., and Hussey, M.J.L. (1967), *The Theory of One-Dimensional Consolidation of Saturated Clays, 1. Finite Non-Linear Consolidation of Thin Homogeneous Layers,* Geotechnique, 17, pp. 261-273.

Gibson, R.E., and Shefford G.C. (1968), *The Efficiency of Horizontal Drainage Layers for Accelerating Consolidation of Clay Embankments,* Geotechnique, 18, pp. 327-335.

Gibson, R.E., Schiffman, R.L., and Cargill, K.W. (1981), *The Theory of One-Dimensional Consolidation of Saturated Clays, II. Finite Nonlinear Consolidation of Thick Homogeneous Layers,* Canadian Geotechnical Journal, 18, pp. 280-293.

Gidley, J.S., and Sack, W.S. (1984), *Environmental Aspects of Wastes Utilization in Construction,* Journal of Environmental Engineering Division, ASCE, 110(6), pp. 1117-1133.

Goh, A.T.C. (1994), *Seismic Liquefaction Potential Assessed by Neural Networks,* Journal of Geotechnical and Geoenvironmental Engineering, ASCE, Vol. 120, No. 9, pp. 1467-1480.

Goh, A.T.C. (1995), *Modeling Soil Correlations Using Neural Networks,* Journal of Computing in Civil Engineering, ASCE, Vol. 9, No. 4, pp. 275-278.

Goldberg, D.E. (1989), *Genetic algorithms in search optimization and machine learning,* Addison-Wesley, Reading, Mass.

Grace, H., and Green, P.A. (1978), *The Use of Wet Fill for the Construction of Embankments for Motorways,* Clay Fills, Institution of Civil Engineers, London, pp. 113-118.

Green, R.G.V., and Hawkins, A.B. (1987), *Assessment of Embankment Suitability,* Compaction Technology, Thomas Telford, London, pp. 91-109.

Greeuw, G. (1997), *Manual FSCONBAG Version 2.1, A Program to Simulate Consolidation of Sludge,* Delft Geotechnics, Delft.

Holland, J.H. (1975), *Adaptation in natural and artifical systems,* University of Michigan Press, Ann Arbor, Mich.

Hornik, K., Stinchcombe, M., and White, H. (1989), *Multilayer feedforward networks are universal approximators,* Neural Networks, 2, pp. 359-366.

Inada, M., Nishinakamura, K., Kondo, T., Shima, H., and Ogawa, N. (1978), *On the Long-term Stability of An Embankment of Soft Cohesive Volcanic Soil,* Clay fills, Institution of Civil Engineers, London, pp. 127-132.

Jacobs, T.L. (1997), *Applications of Optimization Techniques to Structural Design,* Design and Operation of Civil and Environmental Engineering Systems, John Wiley & Sons, New York.

Karshenas, S. (1989), *Truck capacity selection for earthmoving,* J. Constr. Engrg. and Mgmt., ASCE, 115(2), 212-227.

Karunaratne, G.P., Young, K.Y., Tan, T.S., Tan, S.A., Liang, K.M., Lee, S.L., and Vijiaratnam, A. (1990), *Layered Clay-Sand Scheme Reclamation at Changi South Bay,* Tenth Southeast Asian Geotecnnical Conference, Taipei, pp. 71-76.

Koppula, S.D., and Morgenstern, N.R. (1972), *Consolidation of Clay Layer in Two Dimensions,* Journal of the Soil Mechanics and Foundations Division, ASCE, Vol. 98, No. SM1, pp. 79-93.

Krizek, R.J., Karadi, G.M., and Hummel, P.L. (1973), *Engineering Characteristics of Polluted Dredgings,* Technical Report Number 1, Northwestern University, Evanston, Illinois.

Krizek, R.J., Roderick, G.L., and Jin, J.S. (1978), *Use of Dredgings for Landfill, Technical Report Number 2, Stabilization of Dredged Material*, Northweatern University, Evanston, Illinois.

Kruse, G.A.M., and Nieuwenhuis, J.D. (1998), *Impact of Weathering on Erosion Resistance of Cohesive Soil*, 8th International IAEG Congress, Vancouver, pp. 4299-4306.

Kyfor, Z.G., and Gemme, R.L. (1994), *Compacted Clay Embankment Failures*, Transportation Research Record 1462, TRB, National Research Council, Washington, DC, pp. 17-27.

Lee, S.L., Karunaratne, G.P., Yong, K.Y., and Ganeshan, V. (1987), *Layered Clay-Sand Scheme of Land Reclamation*, Journal of Geotechnical Engineering, ASCE, Vol. 113. No. 9, pp. 984-995.

Limsiri, C., Lubking, P., and Kruse, G. (2003), *Optimisation of Earthmoving Equipment Selection by Genetic Algorithms*, Theory and Practice of Modern civil Engineering, Hohai University Press, Nanjing, pp. 301-306.

Lounis, Z., and Cohn, M.Z. (1993), *Multiobjective optimization of prestressed concrete structures*, J. Struct. Engrg., ASCE, 119(3), pp. 794-808.

Maier, H.R., and Dandy, G.C. (2000), *Applications of artificial neural networks to forecasting of surface water quality variables: Issues, applications and challenges*, Artificial neural networks in hydrology, Kluwer, Dordrecht, The Netherlands, pp. 287-309.

Malkawi, A.I.H., Hassan, W.F., and Sarma, S.K. (2001), *Global Search Method for Locating General Slip Surface Using Monte Carlo Techniques*, Journal of Geotechnical and Geoenvironmental Engineering, ASCE, Vol. 127, No. 8, pp. 688-698.

Mateos, M. (1964), *Soil Lime Research at Iowa State University*, Journal of the Soil Mechanics and Foundations Division, ASCE, Vol. 90, No. SM2, pp. 127-153.

Michalewicz, Z. (1992), *Genetic algorithms + data structures = evolution programs*, Springer-Verlag, Berlin.

Michalewicz, Z., and Shoenauer, M. (1996), *Evolutionary Algorithms for Constrained Parameter Optimization Problems*, Evolutionary Computation, Vol. 4, No. 1, pp. 1-32.

Mitchell, J.K. (1976), *Fundamentals of Soil Behavior*, John Wiley & Sons, New York.

Mitchell, M. (1996), *An Introduction to Genetic Algorithms*, The MIT Press.

Najjar, Y.M., and Basheer, I.A. (1996), *A Neural Network approach for Site Characterization and Uncertainty Prediction*, Geotechnical Special Publication, 58(1), pp. 134-148.

Nash, D. (2000), *Modelling the Effects of Surcharge to Reduce Long Term Settlement of Reclamations Over Soft Clays: A Numerical case study*, Soils and Foundations, Vol. 41, No. 5, pp. 1-13.

Onwubiko, C. (2000), *Introduction to Engineering Design Optimization*, Prentice Hall, New Jersey.

Operating cost standards for construction equipment. 11th revised edition, Alphen aan den Rijn, Samsom, 1995.

Ostlid, H. (1981), *High Clay Road Embankments*, Norwegian Road Research Laboratory, Meddelelse nr. 54, pp. 5-18.

Parsons, A.W. (1976), *The Rapid Measurement of the Moisture Condition of Earthwork Material*, TRRL Laboratory Report 750, Transport and Road Research Laboratory, Crowthorne.

Parsons, A.W. (1978), *General report: Construction and Placement of Clay Fills*, Clay Fills, Institution of Civil Engineers, London, pp. 307-314.

Peurifoy, R.L., Ledbetter, W.B., and Schexnayder, C.J. (1996), *Construction planning, equipment, and methods*, McGraw-Hill, New York.

Pezeshk, S., Camp, C.V., and Chen, D. (2000), *Design of nonlinear framed structures using genetic optimization*, J. Struct. Engrg., ASCE, 126(3), 382-388.

Rardin, R.L. (1998), *Optimization in Operations Research*, Prentice Hall, New Jersey.

Riele, J.L.M. te. (1999), *GWW KOSTEN bemalingen, grondwerken, drainage.* 14[e] editie, Elsevier bedrijfsinformatie bv., Doetinchem.

Ritzel, B.J., Eheart, J.W., and Ranjithan, S. (1994) *Using Genetic Algorithms to Solve a Multiple Objective Groundwater Pollution Containment Problem*, Water Resources Research 30, 5 (May), pp 1589-1603.

Rollings, M.P., and Rollings, R.S. Jr. (1996), *Geotechnical Materials in Construction*, McGraw-Hill, New York.

Russell, J.E. (1985), *Construction Equipment*, Prentice-Hall, Virginia.

Russell, S., and Norvig, P. (1995), *Artificial Intelligence: A Modern Approach*, Prentice Hall, Englewood Cliffs, New Jersey.

Sain, C.H., and Quinby, G.W. (1996), *Earthwork*, In: Merritt, F.S., Loftin, M.K., and Ricketts, J.T. (eds.), *Standard Handbook for Civil Engineers*, McGraw-Hill, New York.

Saribas, A., and Erbatur, F. (1996), *Optimization and Sensitivity of Retaining Structures,* Journal of Geotechnical Engineering, ASCE, Vol. 122, No. 8, pp. 649-656.

Scott, C.R. (1980), *Soil Mechanics and Foundations*, Applied Science Publishers, London.

Seneviratne, N.H., Fahey, M., Newson, T.A., and Fujiyasu, Y. (1996), *Numerical Modelling of Consolidation and Evaporation of Slurried Mine Tailings*, International Journal for Numerical and Analytical Methods in Geomechanics, Vol. 20, pp. 647-671.

Shahin, M, A., Maier, H.R., and Jaksa, M, B. (2002), *Predicting Settlement of Shallow Foundations using Neural Networks*, Journal of Geotechnical and Geoenvironmental Engineering, Vol. 128, No. 9, pp. 785-793.

Smith, A.A., Hinton, E., and Lewis, R.W. (1983), *Civil Engineering Systems Analysis and Design*, John Wiley & Sons, New York.

Smith, M. (1993), *Neural networks for statistical modelling*, Van Nostrand-Reinhold, New York.

Stewart, R.D. (1991), *Cost estimating,* John Wiley & Sons, New York.

Swarbrick, G.E., and Fell, R. (1992), *Modelling desiccation of mine tailings*, Journal of Geotechnical Engineering, ASCE, Vol. 118. No. 4, pp. 540-557.

Tan, Siew-Ann., Liang, Kee-Ming., Yong, Kwet-Yew., and Lee, Seng-Lip. (1992), *Drainage Efficiency of Sand Layer in Layered Clay-Sand Reclamation*, Journal of Geotechnical Engineering, ASCE, Vol. 118, No. 2, pp. 209-228.

Teh, C.I., Wong, K.S., Goh, A.T.C., and Jaritngam, S. (1997), *Prediction of Pile Capacity Using Neural Networks*, Journal of Computing in Civil Engineering, ASCE, Vol. 11, No. 2, pp. 129-138.

Thanedar, P.B., and Vanderplaats, G.N. (1995), *Survey of Discrete Variable Optimization for Structural Design*, Journal of Structural Engineering, Vol. 121, No. 2, pp. 301-306.

Thomas, B.R. (1990), *Clyde Sediments: Physical Conditioning in Relation to Use as a Topsoil Product for Land Reclamation*, PhD Thesis, Department of Civil Engineering, University of Strathclyde, Glasgow.

Thomas, B.R. (1993), *Wet Fills: evaporative dewatering techniques applied to earthworks construction*, Engineered fills, Thomas Telford, London, pp 99-108.

Twomey, J.M., and Smith, A.E. (1997), *Validation and verification*, Artificial neural networks for civil engineers: Fundamentals and applications, ASCE, New York, pp. 44-64.

U.S. Army Corps of Engineers (1987), *Engineering and Design " Confined Disposal of Dredged Material"*, Engineering Manual 1110-2-5027, U.S. Army Corps of Engineers, Washington, DC.

Vereniging Grootbedrijf Bouwnijverheid (1995), *Kostennormen voor aannemersmaterieel*, Alphen aan den Rijn, Samsom BedrijfsInformatie, Diegem.

Willet, J.R. (1972), *Bodemfysisch Gedrag van Opgespoten Baggerspecie uit de Rotterdamse Havens (Het Begreppelen van de Rotterdamse Baggerdepots)*, De Ingenieur, Jrg. 84, nr. 7, pp. B1-B12.

Wind, G.P. and Van Doorne, W. (1975), *A Numerical Model for the Simulation of Unsaturated Vertical Flow of Moisture in Soils*, Journal of Hydrology, 24, pp. 1-20.

APPENDIX A - PRODUCTION RATE

A.1   Bulldozer

Production rate = (hp * 330 / (D / 0.3048 + 50)) * 0.7646 * operator factor * material
type * dozing factor * job efficiency * grade factor

* One-way push distance, D = 50 m
* Operator factor = 0.75 for $I_C \geq 0.70$ clay and sand: 0.60 for $0.50 < I_C < 0.70$ clay
    * Excellent = 1
    * Average = 0.75
    * Poor = 0.60
* Material factor = 0.80 for clay: 1.20 for sand
    * Loose stockpile = 1.20
    * Very sticky = 0.80
* Dozing factor = 1.20
    * Slot = 1.20
    * Side by side = 1.15 - 1.25
* Working time
    * 50 min/hr, job efficiency = 0.83 for $I_C \geq 0.70$ clay and sand
    * 40 min/hr, job efficiency = 0.67 for $0.50 < I_C < 0.70$ clay
* Grade = 0 %, grade factor = 1
    * Grade factor = $-7*10^{-5}*(\%grade)^2 - 0.0221*(\%grade) + 1$

Identification number	Horsepower	Production rate $(1\ m^3/hr)$		
		$I_C \geq 0.70$	$0.50 < I_C < 0.70$	Sand
B1	95	67	43	100
B2	135	95	61	143
B3	180	127	81	190
B4	200	141	90	212
B5	280	198	127	297

A.2   Loader

Production rate = ((60/total cycle time) * material factor * bucket capacity) * job
efficiency

The cycle time per load:
* Track loader for $0.50 < I_C < 0.70$ clay
    * Fix time to load, turn, and dump 0.30 min
    * Haul time = one way haul distance / ((0.0144 * hp + 3.69) * 1000/60) min
    * Return time = one way haul distance / ((0.047 * hp + 1.69) * 1000/60) min
    * Distance = 10 m
* Wheel loader for $I_C \geq 0.70$ clay and sand
    * Fix time to load, turn, and dump 0.50 min
    * Haul time = one way haul distance / ((0.1214 * hp + 0.27) * 1000/60) min

- Return time = one way haul distance / ((0.1061 * hp + 9.06) * 1000/60) min
- Distance = 10 m
* Material factor
  - 0.90 for clay
  - 0.975 for sand
* Working time
  - 50 min/hr, Job efficiency = 0.83 for $I_C \geq 0.70$ clay and sand
  - 40 min/hr, job efficiency = 0.67 for $0.50 < I_C < 0.70$ clay

Track loader identification number	Horsepower	Bucket capacity (m^3)	Haul time (min)	Return time (min)	Total cycle time (min)	Production rate $(1\ m^3/hr)$
TL1	70	1	0.13	0.12	0.55	65
TL2	95	1.5	0.12	0.10	0.52	105
TL3	160	2.5	0.10	0.07	0.47	193
TL4	215	3.5	0.09	0.05	0.44	287

Wheel loader identification number	Horsepower	Bucket capacity (m^3)	Haul time (min)	Return time (min)	Total cycle time (min)	Production rate $(1\ m^3/hr)$	
						$I_C \geq 0.70$	Sand
WL1	70	1	0.07	0.04	0.40	111	120
WL2	95	1.5	0.05	0.03	0.38	177	191
WL3	160	2.5	0.03	0.02	0.35	318	345
WL4	215	3.5	0.02	0.02	0.34	461	500

A.3   Hydraulic hoe

Production rate = (60/total cycle time * heaped bucket capacity * bucket fill factor) * job efficiency

* The cycle time per load:
    Total cycle time = 5.76 * Ln(heaped bucket capacity) + 20.68
    (Adapting from Caterpillar, 1995)
* Fill factors for hydraulic hoe buckets:

Material	Fill factor (%)
Moist loam / sandy clay	100 - 110
Sand and gravel	95 - 110
Rock-poorly blasted	40 - 50
Rock-well blasted	60 - 75
Hard, tough clay	80 - 90

* Fill factor = 0.85
* Working time
  - 50 min/hr, Job efficiency = 0.83 for $I_C \geq 0.70$ clay and sand
  - 40 min/hr, job efficiency = 0.67 for $0.50 < I_C < 0.70$ clay

Identification Number	Horsepower	Bucket capacity (m^3)	Total cycle time (min)	Production rate (1 m^3/hr)	
				$I_C \geq 0.70$	$0.50 < I_C < 0.70$
H1	79	0.45	0.27	71	57
H2	99	0.52	0.28	78	63
H3	99	0.68	0.31	94	75
H4	128	0.80	0.32	105	84
H5	153	1.00	0.34	123	99
H6	168	1.10	0.35	132	106
H7	222	1.40	0.38	158	126
H8	286	1.91	0.41	200	160

A.4   Truck

Production rate = heaped capacity * ((1/total round-trip time) * job efficiency)

* The time required for each operation in a round-trip cycle:
  * Loading = heaped capacity of truck / loader production (hr)
  * Lost time in pit and accelerating = 0.1 * capacity weight (ton) – 1 (hr)
  * Travel to the fill = haul distance / haul velocity (hr)
  * *Haul velocity = (hp * mechanical efficiency * 746) / (gross weight, Newton * total resistance)*
  + Mechanical efficiency = 0.8
  + Total resistance = rolling resistance + grade resistance
    - Rolling resistance = ((rolling factor * weight, ton)/(weight, ton * 1000))*100%
      o Rolling factor = 50
      o Earth, compacted and maintained 25 - 35 kg/ton
* Earth, poorly maintained 35 - 50 kg/ton
    - Grade resistance = haul-road grade %
* Total resistance > 0.025 for haul speed
* Total resistance > 0.055 for return speed
  * Dumping, turning and accelerating = (0.1* capacity weight - 1) - 0.5
  * Travel to pit = haul distance / return velocity
* Haul-road grade = 0%
* Working time
  * 50 min/hr, Job efficiency = 0.83 for $I_C \geq 0.70$ clay and sand
  * 40 min/hr, job efficiency = 0.67 for $0.50 < I_C < 0.70$ clay
* Haul distance = 2 km
* Average velocity = maximum haul and return velocity * 0.8

Identification number	Horsepower	Empty weight (ton)	Capacity weight (ton)	Gross weight (ton)	Heaped capacity (m^3)	Loading height (m)
T1	190	20	20	40	10.0	2.55
T2	220	20	25	45	13.7	2.70
T3	280	20	30	50	16.5	2.85
T4	335	25	35	55	20.5	2.93

For $I_C \geq 0.70$ clay

Id. num.	Haul speed (km/hr)	Return speed (km/hr)	Total round-trip time (hr)				Production rate (l m³/hr)			
			Loader 70 hp	Loader 95 hp	Loader 160 hp	Loader 215 hp	Loader 70 hp	Loader 95 hp	Loader 160 hp	Loader 215 hp
T1	16.33	29.69	0.30	0.27	0.25	0.24	27	31	34	35
T2	16.81	34.38	0.34	0.30	0.26	0.25	33	39	44	46
T3	19.25	43.75	0.36	0.30	0.26	0.25	38	46	53	56
T4	19.19	41.88	0.41	0.34	0.29	0.27	42	50	59	63

For sand

Id. num.	Haul speed (km/hr)	Return speed (km/hr)	Total round-trip time (hr)				Production rate (l m³/hr)			
			Loader 70 hp	Loader 95 hp	Loader 160 hp	Loader 215 hp	Loader 70 hp	Loader 95 hp	Loader 160 hp	Loader 215 hp
T1	16.33	29.69	0.30	0.27	0.24	0.23	28	31	34	35
T2	16.81	34.38	0.33	0.29	0.26	0.25	34	39	44	46
T3	19.25	43.75	0.34	0.29	0.26	0.25	40	47	54	57
T4	19.19	41.88	0.40	0.33	0.29	0.27	43	51	60	64

For $0.50 < I_C < 0.70$ clay

Id. num.	Haul speed (km/hr)	Return speed (km/hr)	Total round-trip time (hr)				Production rate (l m³/hr)			
			Loader 70 hp	Loader 95 hp	Loader 160 hp	Loader 215 hp	Loader 70 hp	Loader 95 hp	Loader 160 hp	Loader 215 hp
T1	16.33	29.69	0.37	0.31	0.27	0.25	18	21	25	27
T2	16.81	34.38	0.43	0.35	0.29	0.27	21	26	32	34
T3	19.25	43.75	0.46	0.37	0.29	0.27	24	30	38	41
T4	19.19	41.88	0.54	0.42	0.33	0.30	25	32	41	46

A.5   Compactor

Production rate = ((drum width * average speed * compacted lift thickness) / number of passes) * 1.59
- *Average speed = 2 km/hr*
- *Compacted lift thickness = 150 mm*
- *Number of passes = 4*

Identification number	Horsepower	Drum width (m)	Production rate (l m³/hr)
C1	77	1.22	120
C2	107	1.68	164
C3	145	2.13	207

APPENDIX B - OWNING AND OPERATING COSTS

B.1   Bulldozer

Equipment cost
Ownership period 7 years
Rest value 5% of standard value
Usage 32 week/year
Depreciation + interest = 17.92% of standard value per year, 0.56% per week
Maintenance + repair:
*   dry condition = 14.08% of standard value per year, 0.44% per week
*   wet condition = 28.80% of standard value per year, 0.90% per week

Fuel cost
Fuel consumption 0.04 gal per fwhp-hr = 0.04*3.785 = 0.1514 litre per fwhp-hr
Fuel consumed   = operating factor * hp * 0.1514
                             = engine factor * job efficiency * hp * 0.1514
        where engine factor   = 0.65
                working time   = 50 min/hr, job efficiency = 0.83 for dry condition
                ....   = 40 min/hr, job efficiency = 0.67 for wet condition

Diesel fuel price = 0.8 €/litre

Identification number	Horsepower	Standard value (€)	Equipment cost (€/hr)		Fuel cost (€/hr)		Total (€/hr)	
			Dry	Wet	Dry	Wet	Dry	Wet
B1	95	102,300	26	37	6.23	5.00	32.23	42.00
B2	135	118,200	30	43	8.86	7.10	38.86	50.10
B3	180	150,000	38	55	11.81	9.45	49.81	64.45
B4	200	227,300	57	83	13.12	10.50	70.12	93.50
B5	280	354,600	89	129	18.37	14.70	107.37	143.70

B.2   Loader

Equipment cost
Ownership period 7 years
Rest value 5% of standard value
Usage 32 week/year
Depreciation + interest = 17.92% of standard value per year, 0.56% per week
Maintenance + repair = 16.96% of standard value per year, 0.53% per week

Fuel cost

Fuel consumption 0.04 gal per fwhp-hr = 0.04*3.785 = 0.1514 litre per fwhp-hr

Fuel consumed   = operating factor * hp * 0.1514

= engine factor * job efficiency * hp * 0.1514

where - engine factor   = 0.65

- working time:

- 50 min/hr, job efficiency = 0.83 for $I_C \geq 0.7$ clay and sand
- 40 min/hr, job efficiency = 0.67 for $0.5 < I_C < 0.7$ clay

Diesel fuel price = 0.8 € /litre

For $I_C \geq 0.7$ clay and sand

Identification Number	Bucket capacity (litre)	Horsepower	Standard value (€)	Equipment cost (€/hr)	Fuel cost (€/hr)	Total (€ /hr)
WL1	1000	70	57,000	15.5	4.59	20.09
WL2	1500	95	82,000	22.5	6.23	28.73
WL3	2500	160	102,500	28.0	10.50	38.50
WL4	3500	215	148,000	40.5	14.11	54.61

For $0.5 < I_C < 0.7$ clay

Identification Number	Bucket capacity (litre)	Horsepower	Standard value (€)	Equipment cost (€/hr)	Fuel cost (€/hr)	Total (€ /hr)
TL1	1000	70	57,000	15.5	3.67	19.17
TL2	1500	95	82,000	22.5	4.99	27.49
TL3	2500	160	102,500	28.0	8.40	36.40
TL4	3500	215	148,000	40.5	11.28	51.78

B.3   Hydraulic hoe

Equipment cost

Ownership period 7 years

Rest value 5% of standard value

Usage 32 week/year, 40 hr/week

Depreciation + interest = 17.92% of standard value per year, 0.56% per week

Maintenance + repair = 10.24% of standard value per year, 0.32% per week

Fuel cost

Fuel consumption 0.04 gal per fwhp-hr = 0.04*3.785 = 0.1514 litre per fwhp-hr

Fuel consumed   = operating factor * hp * 0.1514

= engine factor * job efficiency * hp * 0.1514

where - engine factor = 0.65

- working time = 50 min/hr, job efficiency = 0.83

Diesel fuel price = 0.8 € /litre

Identification number	Bucket capacity (m³)	Horsepower	Standard value (€)	Equipment cost (€/hr)	Fuel cost (€/hr)	Total (€/hr)
H1	0.45	79	75,000	16.5	5.18	21.68
H2	0.52	99	86,000	19.0	6.50	25.50
H3	0.68	99	100,000	22.0	6.50	28.50
H4	0.80	128	120,000	26.0	8.40	34.40
H5	1.00	153	136,000	30.0	10.04	40.04
H6	1.10	168	160,000	35.0	11.02	46.02
H7	1.40	222	182,000	40.0	14.56	54.56
H8	1.91	286	218,000	48.0	18.76	66.76

B.4   Truck

Equipment cost
Ownership period 7 years
Rest value 5% of standard value
Usage 32 week/year
Depreciation + interest = 17.92% of standard value per year, 0.56% per week
Maintenance + repair = 15.04% of standard value per year, 0.47% per week

Fuel cost
Fuel consumption 0.04 gal per fwhp-hr = 0.04*3.785 = 0.1514 litre per fwhp-hr
Fuel consumed   = operating factor * hp * 0.1514
                 = engine factor * job efficiency * hp * 0.1514
    where - engine factor = 0.65
          - working time
            - 50 min/hr, job efficiency = 0.83 for dry material
            - 40 min/hr, job efficiency = 0.67 for wet material

Diesel fuel price = 0.8 € /litre

Identification number	Max. capacity weight (ton)	Horse-power	Standard value (€)	Equipment cost (€/hr)	Fuel cost (€/hr)		Total (€/hr)	
					Dry material	Wet material	Dry material	Wet material
T1	20	190	136,500	35.0	12.47	9.97	47.47	44.97
T2	25	220	148,000	38.0	14.43	11.55	52.43	49.55
T3	30	280	227,500	50.5	18.37	14.70	68.87	65.20
T4	35	335	273,000	70.0	21.98	17.58	91.98	87.58

B.5   Compactor

Equipment cost
Ownership period 8 years
Rest value 5% of standard value
Usage 32 week/year
Depreciation + interest = 16.32% of standard value per year, 0.51% per week
Maintenance + repair = 11.20% of standard value per year, 0.35% per week

Fuel cost

Fuel consumption 0.04 gal per fwhp-hr = 0.04*3.785 = 0.1514 litre per fwhp-hr

Fuel consumed   = operating factor * hp * 0.1514

                            = engine factor * job efficiency * hp * 0.1514

        where engine factor = 0.65

                working time = 50 min/hr, job efficiency = 0.83

Diesel fuel price = 0.8 € /litre

Identification number	Max. capacity weight (ton)	Horsepower	Standard value (€)	Equipment cost (€/hr)	Fuel cost (€/hr)	Total (€/hr)
C1	8	77	36,400	9.50	5.05	14.55
C2	10	107	61,400	15.80	7.02	22.82
C3	12	145	75,000	19.50	9.51	29.01

APPENDIX C - NUMERICAL EXAMPLE OF GENETIC ALGORITHMS WITH PENALTY FUNCTION

This section provides a simple GA in order to clarify how GAs and penalty function work. The genetic algorithm and quadratic penalty function used in the following example is based on the description given on section 4.4. The example is to find the minimum value of $f(x) = x^2$. The value of x is between 0 and 31. The constraint is $x \geq 10$.

$$\text{Min } f(x) = x^2 \qquad\qquad (C-1)$$

$$\text{s.t } x \geq 10 \qquad\qquad (C-2)$$

$$0 \leq x \leq 31 \qquad\qquad (C-3)$$

The penalty objective function can be formulated as:

$$\text{Min } f(x) = x^2 + r[x - 10]^2 \qquad\qquad (C-4)$$

where r = penalty parameter (50 is selected for this problem)

The GA parameters are: population number = 6, probability of crossover = 0.7, and probability of mutation = 0.01 (these figures are random but steered by practical experience).
*   Generate solutions:
    *   *Encoding:*
        The problem size is 0-31 so we can represent each possible population members as a binary string of 5 bits;
    *   *Initialisation:*
        With the population of ten, a random population of 10 chromosomes and each chromosome comprises 5 bits is generated;
        Initial population: (01101, 11000, 01000, 10011, 11110, 00011)
*   Evaluate variables:
    Determine the fitness of each chromosome in the population by evaluating $f(x) = x^2$, checking the violation of the constraint and the fitness ratio by:

$$Fitness\ (chromosome\ \ i) = \frac{1}{f(x)(chromosome\ \ i)} \qquad (C-5)$$

$$Fitness\ ratio\ (chromosome\ \ i) = \frac{Fitness\ (chromosome\ \ i)}{\sum\limits_{i=1}^{n} Fitness\ (chromosome\ \ i)} \qquad (C-6)$$

where n = number of chromosomes

Chromosome	Binary string	Decode (x)	Penalty	Fitness	Fitness ratio
1	01101	13	0	0.0059	0.376
2	11000	24	0	0.0017	0.108
3	01000	8	200	0.0038	0.242
4	10011	19	0	0.0028	0.178
5	11110	30	0	0.0011	0.070
6	00011	3	2450	0.0004	0.025

The average of $f(x) = 346.5$

The minimum of $x = 13$ (without violating constraint)

* Reproduction:

Performing roulette wheel selection:

- Choose a random number r between 1 and 1000;
- If $r \leq 376$ select 01101 else if $377 \leq r \leq 485$ select 11000 else if $486 \leq r \leq 727$ select 01000 else if $728 \leq r \leq 905$ select 10011 else if $906 \leq r \leq 975$ select 11110 else select 00011; (size of the intervals are directly related to the fitness of a particular chromosome)
- Repeating 6 times.

Using random numbers, the trial outcomes are 103, 445, 180, 355, 488, and 758.
The selected chromosomes are:

Chromosome	Binary string
1'	01101
2'	11000
3'	01101
4'	01101
5'	01000
6'	10011

* *Crossover:*

- Choose two chromosomes in the new population at random based on probability of crossover;
- Choose a crossover point within the chromosomes at random;
- Exchange the parts of the two chromosomes, then;

Chromosome 1'	01101
Chromosome 2'	11000
Offspring 1	11001
Offspring 2	01100

Chromosome 3'	01101
Chromosome 4'	01101
Offspring 3	01101
Offspring 4	01101

Chromosome 5'	01000
Chromosome 6'	10011
Offspring 5	10000
Offspring 6	01011

* Mutation:

Based on the probability of mutation, bit by bit is examined and flip the correspondent bit if that bit is selected for mutation. Because it has small probability for mutation, thus only one bit is changed for this example.

Offspring	Binary string
1	11000
2	01100
3	01101
4	01101
5	10000
6	01011

* Evaluation variables:
Calculating the fitness and fitness ratio of the new population (off springs of the previous generation become the new chromosomes of this generation):

2nd generation

Chromosome	Binary string	Decode (x)	Penalty	Fitness	Fitness ratio
1	11000	24	0	0.0017	0.052
2	01100	12	0	0.0069	0.212
3	01101	13	0	0.0059	0.181
4	01101	13	0	0.0059	0.181
5	10000	16	0	0.0039	0.120
6	01011	11	0	0.0083	0.255

The average of $f(x) = 239.2$

The minimum of $x = 11$

The procedure will repeat until the stopping criterion is met. After several iterations (depending on the search space complexity), the algorithm will converge towards the correct answer.

APPENDIX D - NUMERICAL EXAMPLE OF ARTIFICIAL NEURAL NETWORK

A simple example will be used to illustrate the ANN algorithm. This example will provide detail about multi-layer feed-forward ANNs with the back-propagation learning to facilitate an understanding of how neural networks work. An ANN model will be build for a XOR (Exclusive "or") problem. This ANN model takes two boolean inputs and outputs the XOR of them. What we want is an ANN that will output 1 if the two inputs are different and 0 otherwise, as shown in Table D-1.

Table D-1. Input and output of XOR problem

Input 1	Input 2	Desired output
0	1	1
0	0	0
1	0	1
1	1	0

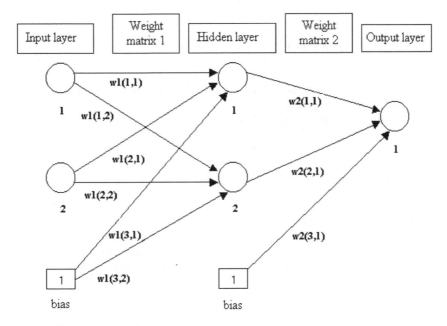

Figure D-1: Five nodes network

A five nodes neuron network, which consists of 2 input nodes, 2 hidden nodes, and 1 output node, is used for this example. The data set for training the network is four as shown in Table G-1. If all values of an input data are zero, the weight in weight matrix would never be changed for this data set and the network could not learn it. Due to that fact, a "pseudo input" called Bias is created. Bias has a constant value of 1. The network is shown in Figure D-1.

Firstly, the network would be initialised, and given random weights, Let's assign these initial weights. The weights can be anything between –1 and 1 as shown in Table D-2.

Table D-2. Initial weights of the network

Node	Weight
Hidden node 1	w1(1,1) = 0.341, w1(2,1) = 0.129, w1(3,1) = -0.923
Hidden node 2	w1(1,2) = -0.115, w1(2,2) = 0.570, w1(3,2) = -0.328
Output node	w2(1,1) = -0.993, w2(2,1) = 0.164, w2(3,1) = 0.752

Since back-propagation and training requires thousands of steps, we are obviously not going to go through it all. We will only look at the first iteration that occurs. Firstly, the sum has to be calculated, then run through the sigmoid function.

In the first training the input value for input node 1 is 0 and the input value for input node 2 is 1. The weight sums at hidden layer are:

Hidden node 1: (0*0.341)+(1*0.129)+(1*-0.923) = -0.794

Hidden node 2: (0*-0.115)+(1*0.570)+(1*-0.328) = 0.242

Pass the weight sum results to the activation function (sigmoid function), which computes the output value of each node.

$$\text{Output of hidden node 1: } \frac{1}{1+e^{-(-0.794)}} = 0.311$$

$$\text{Output of hidden node 2: } \frac{1}{1+e^{-0.242}} = 0.560$$

Using the outputs from the hidden layer as the inputs for the output layer we can make the following calculations:

The weight sum of the output layer is:

Output node: (0.311*-0.993)+(0.560*0.164)+(1*0.752) = 0.535

$$\text{Therefore, the output result} = \frac{1}{1+e^{-0.535}} = 0.631$$

This is the value that the network would output. This value is different from the true value (1 in this case). Thus, we have to adjust all the weights to get the result from the network closer to 1. The back-propagation algorithm is used to do this task. At first we have to calculate the error value by subtracting the output from the target:

Output error = 1-0.631 = 0.369

From the output error we can change the weights in the weight matrix 2:

From Eq. 4.28:

$\Delta_i = (1-0.631)*0.631*(1-0.631) \quad = 0.086$

From Eq. 4.26:

w2(1,1) = -0.993+(1*0.311*0.086) = -0.966

w2(2,1) = 0.164+(1*0.560*0.086) = 0.212

w2(3,1) = 0.752+(1*1*0.086) = 0.838

The weight matrix 1 can be changed as follows:

From Eq. 4.30:

$\Delta_j(1) = 0.311*(1-0.311)*-0.993*0.086 = -0.018$

$\Delta_j(2) = 0.560*(1-0.560)*0.164*0.086 = 0.003$

From Eq. 4.29:

$$w1(1,1) = 0.341 + (1*0*-0.018) \qquad = 0.341$$
$$w1(2,1) = 0.129 + (1*1*-0.018) \qquad = 0.111$$
$$w1(3,1) = -0.923 + (1*1*-0.018) \qquad = -0.941$$
$$w1(1,2) = -0.115 + (1*0*0.003) \qquad = -0.115$$
$$w1(2,2) = 0.570 + (1*1*0.003) \qquad = 0.573$$
$$w1(3,2) = -0.328 + (1*1*0.003) \qquad = -0.325$$

The first input data set had been propagated through the network. The same procedure is used for the next input data set (second line in Table D-1) but then with the changed weight values. After the forward and backward propagation of the fourth data set, one learning step is complete and the network error can be calculated by means of error or root-mean-square error. By performing this procedure repeatedly, this error value gets smaller and smaller. The algorithm is successfully finished, if the network error is zero or approximately zero. For this example the network could balance out the weights to the values as shown in the Table D-3.

Table D-3. Final weights of the network

Node	Weight
Hidden node 1	$w1(1,1) = -7.194$, $w1(2,1) = -7.183$, $w1(3,1) = 3.086$
Hidden node 2	$w1(1,2) = -5.757$, $w1(2,2) = -5.753$, $w1(3,2) = 8.619$
Output node	$w2(1,1) = -12.400$, $w2(2,1) = 12.310$, $w2(3,1) = -5.955$

With these results, we can calculate the results for XOR as follows:

0 XOR 1 = 0.996
0 XOR 0 = 0.004
1 XOR 0 = 0.996
1 XOR 1 = 0.005

Which, with a small amount of rounding, is the same amount as the data that we had trained the network.

APPENDIX E - FSCONBAG SIMULATION

E.1   General

FSCONBAG is a computer program, which models the consolidation behaviour of sludge layers. The program FSCONBAG 2.1 is the product of research and development at Delft Geotechnics (GeoDelft at present) with feedback from Dutch Public Works (DWW). The numerical algorithm in FSCONBAG is the finite difference. The MS-DOS program is written in the programming language Pascal.

E.2   Theoretical background

The FSCONBAG's consolidation model has been based on the so-called Finite Strain Theory (FST), proposed in Gibson (1967) and extended in Gibson (1981). This theory has some advantages over the well-known classical Terzaghi theory. Disadvantages of the Terzaghi theory are the usual restrictions of small strains, a linear stress-strain relationship, and permeability constant over the depth of the layer. In FST special reduced coordinates are employed to avoid the assumption of small deformation in Terzaghi theory. The advantage is that deformations are not limited to small values. Secondly, the material behaviour is not described by a single consolidation coefficient c_v, but depends on the void ratio e. Thus, the material behaviour changes during the consolidation process. A disadvantage of FST is the higher complexity and the non-linearity of the governing equation, which is given below:

$$-\frac{\partial e}{\partial t} = \left[\frac{\gamma_s}{\gamma_w} - 1\right]\frac{d}{de}\left[\frac{k(e)}{1+e}\right]\frac{\partial e}{\partial z} + \frac{\partial}{\partial z}\left[\frac{k(e)}{\gamma_w(1+e)}\frac{d\sigma'}{de}\frac{\partial e}{\partial z}\right] \qquad \text{(E-1)}$$

where e = void ratio
     t = time (day)
     γ_s = solid density (kN/m^3)
     γ_w = water density (kN/m^3)
     k = permeability (m/s)
     z = reduced verticle material coordinate (m)
     σ = effective stress (kPa)

This equation is suitable for large strains, a non-linear stress-strain relationship, and variable permeability. It has been shown to be the most general and least restriction equation describing one-dimensional primary consolidation (Cargill, 1984). For a complete theoretical derivation, the reader is referred to Gibson et al (1967). However, Eq. E-1 can be understood in a general way by examination of the term. The change of the void ratio to time is caused by two components. The first is the action of self-weight which represents by the first term. The second term is similar to the right hand side of Terzaghi's equation.

$$\frac{\partial u}{\partial t} = c_v \frac{\partial^2 u}{\partial z^2} \tag{E-2}$$

where u = excess pore pressure

We note that the coefficient c_v related to:

$$\frac{k(e)}{\gamma_w(1+e)} \frac{d\sigma'}{de} \tag{E-3}$$

We can see that the permeability $k(e)$ and the effective stress $\sigma'(e)$ are functions of the void ratio e. Given these relations there is no need for a constant c_v. It also can say that Terzaghi's equation is a special case of the Finite Strain equation, found by neglecting self-weight and assuming that the production in expression E-3 is not depend on e.

E.3.   Parameter description

FSCONBAG needs material properties for input material parameters. Some of them are straightforward, like the specific gravity of soil and water, and will not be discuss here. The attention will pay on the significant parameters as follows:
*   *The initial void ratio* (e_0)
     The initial void ratio is defined as the void ratio at the transition from sedimentation to consolidation. This value can be estimated in laboratory by a small column test. The empirically determination suggested by U. S. Army Corps of Engineers (1987) is the void ration at a water content of approximately 1.8 times the Atterberg liquid limit of the material;
*   *The relationship between void ratio and effective stress*
     The relation $\sigma'(e)$ is an analytical function that approximates the relation between the effective stress σ' and the void ratio e during virgin compression. Such a function can only be an approximation, since it is known that effective stress also depends on time and chemical interactions. However, Finite Strain Theory assumes that the void ration dependence is by far more important than all other influences. In FSCONBAG the following two functions are implemented:

HYDEX:

$$\sigma'(e) = \sigma_0 [\exp(m_1 + m_2 \cdot e) + \exp(m_3 + m_4 \cdot e)] \tag{E-4}$$

DIRECT:

$$\sigma'(e) = \sigma_0 \cdot \exp(m_1 + m_2 \cdot e + m_3 \cdot e^2 + m_4 \cdot e^3) \tag{E-5}$$

where m_1-m_4 = sludge dependent parameters
      σ_0 = 1 kPa

The HYDEX double exponential function has been derived from HYDCON (the test procedure is given in manual of FSCONBAG version 2.1) and oedometer tests.

To find the parameters, finite strain modelling on the HYDCON data is necessary. The DIRECT function is obtained by polynomial fitting on the HYDCON and oedometer tests. In 1995 a database with sludge data from HYDCON tests and classification tests has been constructed. In the database standard soil classification data are compared with the sludge parameters mentioned above. A relevant part of the database has been incorporated in FSCONBAG. This means that the consolidation parameters and the void ratio can be obtained from the soil classification data.

* *The relationship between void ratio and permeability*
The relation $k(e)$ is approximated by an analytical function too and gives the dependence of the hydraulic permeability k on the void ratio e. However other factors, like temperature and salt concentration, may change $k(e)$; these factors are not modelled. In FSCONBAG the following exponential relation is assumed:

$$k(e) = k_0 \cdot \exp\left(n_1 + n_2 \cdot e + n_3 \cdot e^2\right) \tag{E-6}$$

where n_1-n_3 = sludge dependent parameters
           k_0 = 1 m/s

A HYDCON test or a database is needed to find the parameters. The database that mention above also provides n_1-n_4 for difference type of soil and incorporated in FSCONBAG.

E.4 The result

In this study, FSCONBAG was used to simulate the consolidation behaviour of very soft clay. A total of 282 FSCONBAG analysis runs were conducted with a one clay layer model by varying a number of parameters. These parameters are plasticity index varying from 40 to 80%, initial void ratio varying from 1 to 3, layer thickness varying from 0.3 to 3.8m, and applied load varying from 10 to 200 kPa. At the end of each analysis, total settlement and settlement at different degree of consolidation were recorded. The results from these analyses were used to form the database for developing the ANN model for total settlement (282 data sets) and the ANN model for time of settlement (1085 data sets).

* *Database for developing total settlement ANN model*
There are 282 data sets in this database. One data set consists of 5 items, which are plasticity index, void ratio, clay thickness, surcharge load, and settlement amount. A sample of the data set is shown as follows:

PI (%)	VOID	THICK (m)	SUR (kPa)	SET (m)
80	2	1	150	0.247

* *Database for developing settlement time ANN model*
There are 1085 data sets in this database. One data set consists of 6 items, which are plasticity index, void ratio, clay thickness, surcharge load, settlement amount, and settlement time. A sample of the data set is shown as follows:

PI (%)	VOID	THICK (m)	SUR (kPa)	SET (m)	TIME (day)
40	2	1	50	0.29	37

APPENDIX F - MStab SIMULATION

F.1   General

MStab is a computer program, which is used to perform slope stability analysis. The program MStab is the product of research and development at GeoDelft. MStab version 9.7 has been used in this study.

F.2   Input data

*   *Soil profile*
     Two-dimensional embankment structure is composed of several clay and sand layers with varying thickness.
*   *Loads*
     Distributed loads are positioned as surcharge load on the surface of the embankment structure.
*   *Strength parameter*
     Soil properties are defined for clay and sand. Strength parameter for sand is internal friction angle ϕ and for clay is undrained shear strength C_u.
*   *Calculation method*
     Bishop method is applied for slope stability analysis. The program determines automatically the lowest safety factor, assuming circular shaped slip planes and vertical force equilibrium.

F.3   Results

MStab was used to analyse slope stability of clay-sand layered embankment. A total of 514 MStab analysis runs were conducted with a clay-sand layered model by varying a number of parameters. These parameters are undrained shear strength, clay layer thickness, number of sand layers, and embankment height. The undrained shear strength is varied from 30 to 80 kPa, the clay layer thickness is varied from 0.3 to 3.8m, the number of sand layers is varied from 1 to 16 layers, and the embankment height is varied from 0.5 to 8 m. The criteria consists of a constant sand layers thickness 0.20 m, a constant embankment side slope 2h: 1v, the slip circle occurs only in the embankment fill and goes through the toe. At the end of each analysis, the surcharge load which yields F.S.=2 for the embankment was recorded. The results from these analyses (surcharge loads were converted to surcharge height) were used to form the database for developing the ANN model for clay-sand layered embankment stability (514 data sets). The sample of this database is shown as follows:
*   *Database for developing embankment stability ANN model*
     This database consists of 514 data sets. One data set consists of 4 items, which are undrained shear strength (S_u), clay layer thickness (*Thick*), embankment height (*Height*), and surcharge load (*Sur*). A sample of the data set is shown as follows:

S_u (kPa)	Thick (m)	Height (m)	Sur (kPa)
55	0.8	7	54

APPENDIX G - EVALUATION OF HIDDEN LAYER NODES

G.1   Total settlement ANN model

*   *Performance of ANN models with different hidden layer nodes evaluate by coefficient of determination*

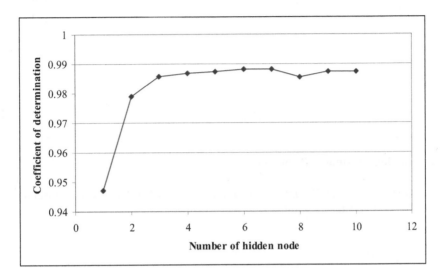

*   *Performance of ANN models with different hidden layer nodes evaluate by root-mean-square error*

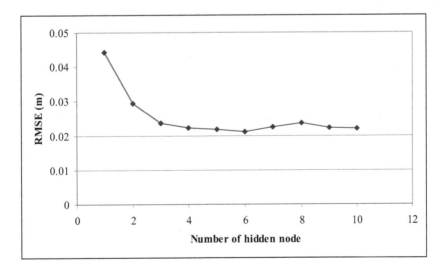

*   *Performance of ANN models with different hidden layer nodes evaluate by mean absolute error*

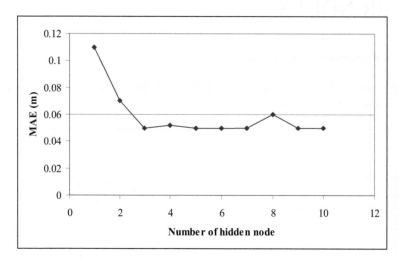

G.2   Settlement time ANN model

*   *Performance of ANN models with different hidden layer nodes evaluate by coefficient of determination*

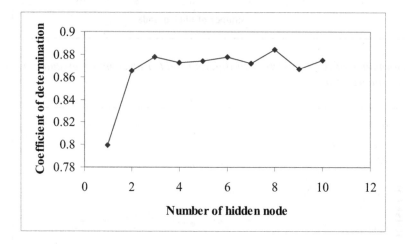

* *Performance of ANN models with different hidden layer nodes evaluate by root-mean-square error*

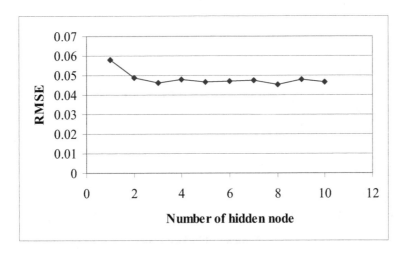

* *Performance of ANN models with different hidden layer nodes evaluate by mean absolute error*

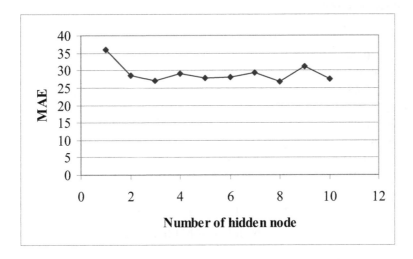

G 3.   Embankment stability ANN model

*   *Performance of ANN models with different hidden layer nodes evaluate by coefficient of determination*

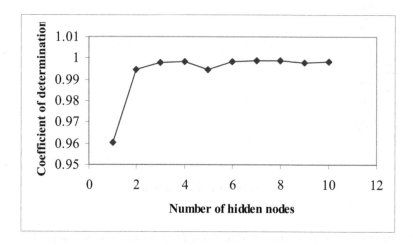

*   *Performance of ANN models with different hidden layer nodes evaluate by root-mean-square error*

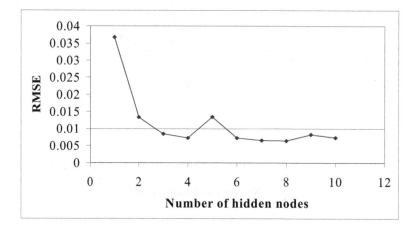

*   *Performance of ANN models with different hidden layer nodes evaluate by mean absolute error*

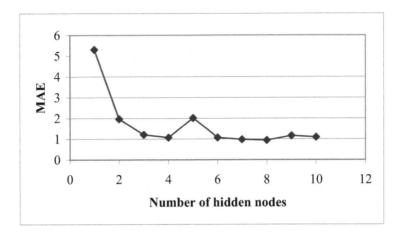

APPENDIX H - SELECTION OF GENETIC ALGORITHMS' PARAMETERS

In this section the selection of the population size, the probability of crossover (p_c), the probability of mutation (p_m), and number of generation for GA parameters of equipment selection model and clay-sand layered embankment model are described. The effects of these parameters on the behaviour of GA are investigated. The investigation is based on the suggested values that were given in GA literatures such as p_c value is from 0.6 to 0.9 (Mitchell, 1996) and a lower bound of p_m value is 1/lenght of chromosome (Back et al., 2000). The variations of the parameters for the investigation are as follows:

* Size of the population: 1.5, 2 and 2.5 time the length of each chromosome
* p_c: 0.6, 0.7, 0.8, and 0.9
* p_m: 0.007, 0.05, 0.1 for equipment selection and 0.03, 0.05, 0.1 for clay-sand layered

The investigation is based on population size = 2 time the length of each chromosome, $p_c = 0.8$, and $p_m = 1$/lenght of chromosome if the parameters are not specified. The number of generation = 500 for equipment selection model and 300 for clay-sand layered model. The curve in the figures for each given parameter is chosen as the maximum value of fitness for five runs of the GA, in which one run is defined as the complete run of GA.

H.1   Selection of GA parameters for equipment selection model

The optimisation formulation of equipment selection model can be seen in section 5.4.1. One chromosome consists of 144-bit string. Number of loaders, bulldozers, and compactors for each model (N_{ij}) is represented by 3-bit string. Number of trucks for each model is represented by 5-bit string. The discrete value of selecting or not selecting of equipment model j type i is represented by 1-bit string. Figure H-1 shows the effects of the population size on the value of the objective function. The size of the population is set to 1.5, 2, and 2.5 times the length of each chromosome. Thus, the population sizes are 216, 288, and 360. The probability of crossover = 0.8, probability of mutation = 0.007 are used in this analysis. The size of the population has no big influence in this case all the populations can reach the optimum result but a population size of 360 gives more consistency result, thus it is chosen for further analysis. The effect of p_c and p_m are shown in Figure H-2 and H-3, respectively. The population size of 360 is used for those investigations. Figure H-2 shows that pc of 0.6 cannot reach the global optimum result, it seem that it can find only local minimum not global minimum. For others p_c, they can find the same optimum result but $p_c = 0.8$ and 0.9 give more consistency result than $p_c = 0.7$. Figure H-3 shows that $p_m = 0.007$ gives the best result for this case. Thus, population size of 360, $p_c = 0.8$, and $p_m = 0.007$ are selected for the GA parameters of the equipment selection optimisation model. For the selected parameters the optimum result can be obtained within 400 generations. Therefore, the maximum generation number for stopping running the GA is chosen as 400.

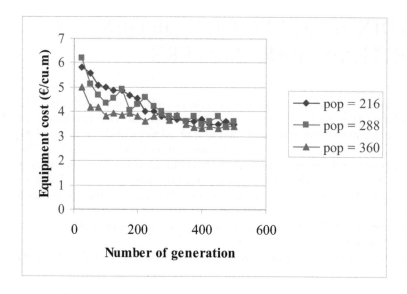

Figure H-1: Relationship between equipment cost and number of generation

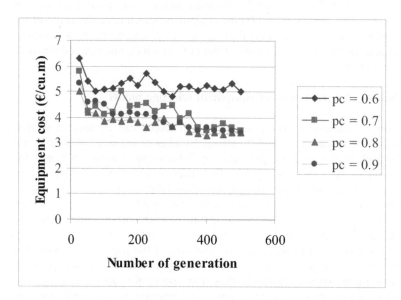

Figure H-2: Effects of crossover rate on the performance of the GA applied to equipment selection model

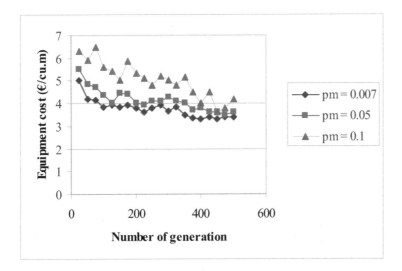

Figure H-3: Effects of mutation rate on the performance of the GA applied to equipment selection model

H.2   Selection of parameters of clay-sand layered model

The GA as described in section 4.4 is applied on the clay-sand layered embankment optimisation model in section 5.2. For this GA, one chromosome consists of 33-bit string. The water content after drying is represented by 6-bit string. The number of clay layers is represented by 4-bit string. The thickness of clay layer is represented by 9-bit string. The surcharge thickness is represented by 8-bit string. The degree of consolidation is represented by 6-bit string. Thus, the variations of the population sizes are 50, 66, and 84. The results of parameter variation are given on Figure H-4 – H-6.

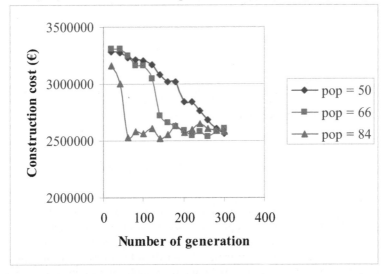

Figure H-4: Relationship between construction cost and number of generation for clay-sand layered model

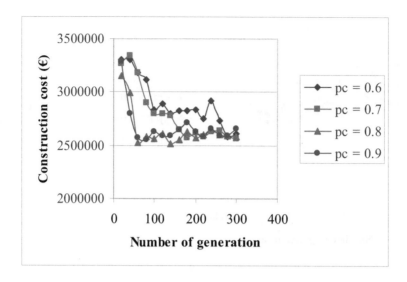

Figure H-5: Effects of crossover rate on the performance of the GA applied to clay-sand layered model

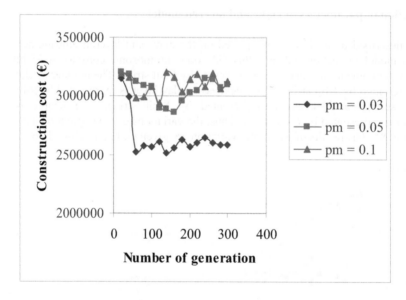

Figure H-6: Effects of mutation rate on the performance of the GA applied to clay-sand layered model

In this case all the population sizes lead to a convergence of the calculation. However, the speed of convergence for population size 84 is faster than the other two population sizes as we can see in Figure H-4. All p_c can also converge to the optimum solution and $p_c = 0.8$ shows the best performance as shown in Figure H-5. For mutation rate, $p_m = 0.03$ give a better result than $p_m = 0.05$ and 0.1 which shows non-consistency result as shown in Figure H-6. Therefore, the population size of 84, $p_c = 0.8$, and $p_m = 0.03$ are selected to be the GA parameters for clay-sand layered embankment model. The maximum generation number for stopping running the GA is chosen as 100.

APPENDIX I - RIJPING SIMULATION

I.1   General

RIJPING is a simulation model for calculation of water balance, ripening, crack formation and surface subsidence of clay sludge. Program RIJPING is an adaptation of program FLOCR that has been described together with some application in Wind and Van Doorne (1975), Bronswijk (1988, 1989).

I.2   Theoretical background

RIJPING is a one-dimensional model that calculates the water balance of clay sludge and outputs for each soil layer are pressure head, water content, crack volume and layer thickness. The effect of weather conditions (rainfall and evaporation), consolidation fluxes, drainage conditions, the occurrence of permanent cracks and changing soil physical characteristics when ripening proceeds have been included. The model compute one-dimensional transient moisture flow based on the combination of Darcy equation and the continuity equation as the following equation:

$$\frac{\partial \theta}{\partial t} = \frac{\partial}{\partial z}\left[k(h)\left(\frac{\partial h}{\partial z} + 1 \right) \right]$$

(I-1)

where θ = volumetric water content
             t = time (d)
             h = pressure head (cm)
             z = distance (cm)
           $k(h)$ = Hydraulic conductivity (cm/d)

In this equation the hydraulic conductivity has been related to the soil metric water pressure (h) through an exponential function. The shrinkage characteristic of a soil has been defined as a relationship between the moisture ration (volume of water over volume of solids) and the void ratio (volume of voids over volume of solids). This shrinkage characteristic is introduced as a third soil-water function besides the water retention curve and the hydraulic conductivity curve. By introducing shrinkage characteristic in simulation model, a clay soil may be considered a continuously changing configuration of soil matrix and shrinkage cracks (Bronswijk, 1988). Subsidence and cracking can be calculated from:

$$V_a = \left(1 + \Delta\right)^r$$

(I-2)

$$V_{cr} = 1 - V_a + \Delta$$

(I-3)

where V_a = actual relative volume

           V_{cr} = actual relative crack volume

           Δ = relative subsidence of a certain soil layer
           r = geometry factor for anisotropic shrinkage

For three-dimensional isotropic shrinkage, r is equal to 3, for one-dimensional subsidence r is equal to 1. Factor r depends on soil texture and sedimentation, ripening stage, moisture content and load. For most ripened soils, $r = 3$ (Boels and Oostindië, 1991). In this model when rainfall exceeds maximum infiltration rate of soil matrix, water flows into cracks. Surface runoff only occurs when cracks are closed. Matrix infiltration and crack infiltration at a given rainfall intensity can be calculated as follows (Figure I-1):

$$P < I_{max}: \qquad I = A_m * P$$

$$I_c = A_c * P$$

$$P > I_{max}: \qquad I = A_m * I_{max}$$

$$I_{c,1} = A_m * (P-I_{max})$$

$$I_{c,2} = A_c * P$$

$$I_c = I_{c,1} + I_{c,2}$$

(I-4)

where: P = Rainfall intensity (cm/day)
I_{max} = Maximum infiltration rate of soil matrix (cm/day)
I = Infiltration rate in soil matrix (cm/day)
I_c = Infiltration rate in cracks (cm/day)
A_m, A_c = Relative areas of soil matrix and cracks respectively
$I_{c,1}$ = Part of total crack infiltration caused by rainfall intensity exceeding maximum infiltration rate of soil matrix (cm/day)
$I_{c,2}$ = Part of total crack infiltration caused by rainfall directly into the cracks (cm/day)

Calculation according to equation I-4, the amount of water infiltration in the cracks depends on rainfall intensity, maximum infiltration capacity of the soil matrix of the top layer and relative are of the cracks. All water infiltrating into cracks is assumed to accumulate at the bottom of the cracks and is added to the moisture content of the corresponding soil layers. Enhancement of soil evaporation because of evaporation out of cracks was not taken into account.

I.3   Input and output

Program RIJPING used a number of input data that consists of the description of soil layers, drainage condition and drainage depth, soil physical characteristics of each soil type, precipitation and evaporation data, and consolidation flux. Water balance, water contents, pressure heads, crack volumes, water content and volume change for the whole soil profile can be obtained from the calculation.

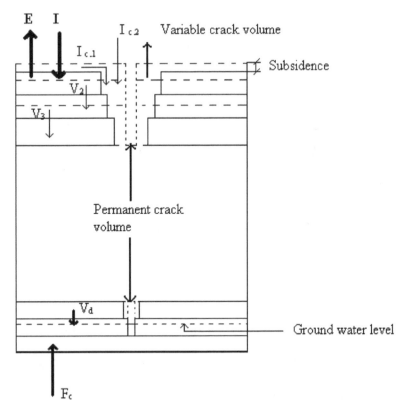

where
I = infiltration rate in soil matrix (cm/d)
$I_{c,1}$ = part of total infiltration caused by rainfall intensity exceeding maximum infiltration rate of soil matrix (cm/d)
$I_{c,2}$ = part of total infiltration caused by direct entering into cracks (cm/d)
E = actual evapotranspiration (m/s)
V = Darcy flux between two nodal points (cm/d)
V_d = flux to groundwater level ($\sim$ drain discharge) (cm/d)
F_c = consolidation flux (cm/d)

Figure I-1: Schematic representation of the simulation model RIJPING

APPENDIX J - DETERMINATION OF COST FUNCTIONS

Cost functions are significant parameters in the optimisation model. The construction cost depends on these parameters. The determination of these parameters is described as follows:

$$\text{Total construction cost} = \text{direct cost} + \text{indirect cost} \qquad \text{(J-1)}$$

The direct cost function comprises four components:
* Cost of clay;
* Cost of horizontal sand layers;
* Cost of surcharge load;
* Cost of sand bunds.

Cost of clay

$$\text{Cost of clay} = \text{unit cost of clay} * \text{volume of clay} \qquad \text{(J-2)}$$

* Unit cost of clay

$$\text{Unit cost of clay} = \text{unit cost of conditioning} + \text{unit cost of construction} \qquad \text{(J-3)}$$

* Unit cost of conditioning

According to the filed data of Sandwich project at Slufter, an excavator agitated the clay every 1-2 week. The average direct cost for an excavator is 0.33 €/m^3. If we take an average of 10 days/agitation, the agitation cost is 0.03 €/m^3/day. Thus:

$$\text{Drying cost} = 0.03 * t_d \qquad \text{(J-4)}$$

where t_d = drying time (day)

The relationship between the drying time and water content in the clay which is expressed using the ratio of the water content of clay after drying (w_d) to the initial water content of clay before drying (w_{in}) can be determine from the data collected from the drying field at Slufter during June 2002 to August 2002. This relationship is shown in Figure J-1and the relationship can be written as:

$$t_d = -\frac{\left(\dfrac{w_d}{w_i}\right)}{0.0046} + 217.40 \qquad \text{(J-5)}$$

w_i = initial water content

Thus:

$$\text{Drying cost} = -6.52 * \left(\frac{w_d}{w_i} \right) + 6.52 \tag{J-6}$$

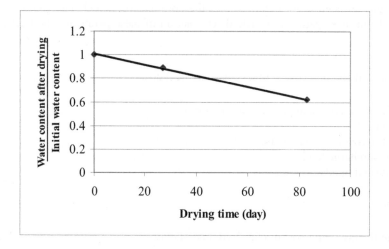

Figure J-1: The relationship between drying time and water content of the clay

Drying cost is the unit cost of conditioning.

*   Unit cost of construction

The unit cost of construction depends on equipment and haul distance. The relationship between the optimum unit cost of construction (direct cost) and haul distance can be determined with the same approach as describe in section 5.4. The relationship is shown in Figure J-2 and J-3. Thus:

For $I_C > 0.7$ clay

$$\text{Unit cost of construction} = 0.68 * d_h + 1.40 \tag{J-7}$$

For $0.5 < I_C < 0.7$ clay

$$\text{Unit cost of construction} = 0.61 * d_h + 2.40 \tag{J-8}$$

These costs are given for the clay at the water content at delivery. Volume decrease due to drying prior to construction is not considered.

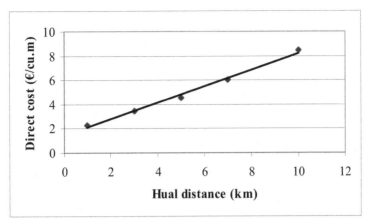

Figure J-2: The relationship between haul distance and direct cost for $I_C > 0.7$ clay

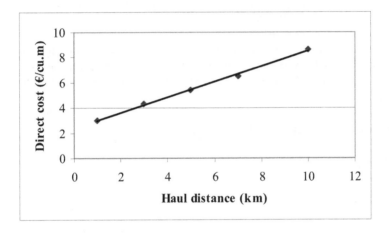

Figure J-3: The relationship between haul distance and direct cost for $0.5 < I_C < 0.7$ clay

where d_h = haul distance (km)

*   Volume of clay

Volume of clay = number of clay layers * average volume of clay layers   (J-9)

For $I_C > 0.7$ clay

$$Volume\ of\ clay = (N_c * C_{in}) * (W_r + 2*(N_c*(C_{in} + S_h))) * L_r \qquad (J\text{-}10)$$

For $0.5 < I_C < 0.7$ clay

$$Volume\ of\ clay = (N_c * C_{in}) * (W_r - 2*(N_c*(C_{in} + S_h))) * L_r \qquad (J\text{-}11)$$

where N_c = number of clay layers
              C_{in} = initial clay thickness (m)

W_r = road width (m)
S_h = horizontal sand thickness (m)
L_r = road length (m)

Cost of horizontal sand layers

Cost of sand layers = unit cost of sand * volume of horizontal sand layer (J-12)

*   Unit cost of horizontal sand layers

Unit cost of sand = unit cost of material + unit cost of construction   (J-13)

- Unit cost of material
  The material cost of sand can be obtained from GWW KOSTEN, bemalingen, grondwerken, drainage 14^e editie, 1999. The unit cost of horizontal sand layers that is drain-sand is 10 €/m^3.
- Unit cost of construction
  The relationship between the optimum unit cost of construction (direct cost) of horizontal sand layers and haul distance can be determined with the same approach as describe in section 5.4. The relationship is shown in Figure J-4. Thus:

Unit cost of construction = 0.54 * d_h + 1.52   (J-14)

where d_h = haul distance

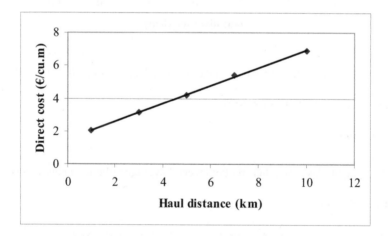

Figure J-4: The relationship between haul distance and direct cost for horizontal sand layers

*   Volume of horizontal sand layers

Volume of sand = number of sand layers * average volume of sand layers(J-15)

For $I_C > 0.7$ clay

$$\text{Volume of sand} = (N_s * S_h) * (W_r + 2 * (N_s * (C_{in} + S_h))) * L_r \qquad \text{(J-16)}$$

For $0.5 < I_C < 0.7$ clay

$$\text{Volume of sand} = (N_s * S_h) * (W_r - 2 * (N_s * (C_{in} + S_h))) * L_r \qquad \text{(J-17)}$$

where N_s = number of horizontal sand layers
C_{in} = initial clay thickness (m)
W_r = road width (m)
S_h = horizontal sand thickness (m)
L_r = road length (m)

Cost of surcharge load

Cost of surcharge load = unit cost of surcharge * volume of surcharge   (J-18)

*   Unit cost of surcharge

Unit cost of surcharge = unit cost of material + unit cost of construction   (J-19)

- Unit cost of material
  The material used for surcharging is street-sand. The material cost of the sand can be obtained from GWW KOSTEN, bemalingen, grondwerken, drainage 14^e editie, 1999. The unit cost of sand for surcharging is 5 €/m^3.
- Unit cost of construction
  The relationship between the optimum unit cost of construction (direct cost) of surcharge load and haul distance is the same as the relationship of horizontal sand layers and haul distance.

It is assumed that after completion of the surcharging that the surcharge load can be sold at the same unit price, as it was bough. Thus, the cost of surcharge load can be determined as:

$$\text{Cost of surcharge load} = (0.54 * d_h + 1.52) + 5 * (1 + r)^{t_w} - 5 \qquad \text{(J-20)}$$

where r = rate of return per day (%)
t_w = the waiting time for surcharging (day)
d_h = haul distance

*   Volume of surcharge load

$$\text{Volume of surcharge load} = S * (W_r - 2 * S) * L_r \qquad \text{(J-21)}$$

where S = the thickness of surcharge load (m)
W_r = road width (m)
L_r = road length (m)

Cost of sand bunds

Cost of sand bunds = unit cost of sand * volume of sand bunds   (J-22)

* Unit cost of sand

Unit cost of sand = unit cost of material + unit cost of construction   (J-23)

- Unit cost of material
  The material used for sand bunds is street-sand. The material cost of the sand can be obtained from GWW KOSTEN, bemalingen, grondwerken, drainage 14^e editie, 1999. The unit cost of sand for sand bunds is 5 €/m^3.
- Unit cost of construction
  The relationship between the optimum unit cost of construction (direct cost) of sand bunds and haul distance is the same as the relationship of horizontal sand layers and haul distance.

* Volume of sand bunds

For $0.5 < I_C < 0.7$ clay

$$Volume\ of\ sand\ bunds = 4*(N_c*(C_{in}+S_h))^2*L_r$$ (J-24)

where N_c = number of clay layers
          C_{in} = initial clay thickness (m)
          S_h = horizontal sand thickness (m)
          L_r = road length (m)

About the Author

Cheevin Limsiri was born in Bangkok, Thailand on October 20, 1960. He obtained his B.Eng. in civil engineering in 1982 from Kasetsart University, Thailand. In the same year, he joined the Port Authority of Thailand as a civil engineer where he took responsibility in planning and design of port infrastructure. In 1987, he was awarded the Master of Engineering Degree in civil engineering from Chulalongkorn University, Thailand. The topic of his M.Eng. thesis was related to the analysis of designing of mid-stream dolphin on the Chao Praya River. In 1991, he acquired his Post Graduate Diploma with distinction in hydraulic engineering from the International Institute for Hydraulic and Environmental Engineering (IHE) in Delft, the Netherlands.

In 1995, he joined Vongchavalitkul University, Thailand as Head of the Civil Engineering Branch. He lectured hydraulic engineering, highway engineering and foundation engineering courses. During the period 2000-2007, he has undertaken the Ph.D. research supervised by UNESCO-IHE, Institute for water education and Delft University of Technology, the Netherlands. During this period, the first four years were spent in GeoDelft and UNESCO-IHE, and the last three years were spent at Vongchavalitkul University where he has been the vice president for research affaires. His thesis entitled: "Very Soft Organic Clay Applied for Road Embankment: Modelling and Optimisation Approach" has studied strategies for using very soft organic clay as a fill material for road embankment construction. An optimisation technique was applied as a method for using very soft organic clay for road embankments leading to an economic construction. The thesis also developed a rational method and a guideline for conditioning and placing very soft organic clay as a fill material for road embankment.

Since graduation, he has spent 25 years working and researching in the civil engineering field including: hydraulic engineering, highway engineering, geotechnical engineering and optimisation for civil engineering. From 2000 to 2007, Mr. Cheevin Limsiri published two journal and seven symposium papers, most of which were related to his Ph.D. topics.

T - #0077 - 071024 - C166 - 254/178/18 - PB - 9780415384872 - Gloss Lamination